The Roof at the Bottom of the World

The Roof
at the Bottom
of the World

Discovering the

Transantarctic

Mountains

Edmund Stump

Yale UNIVERSITY PRESS

NEW HAVEN AND LONDON

Frontispiece: Atop the Tusk, a 600-foot horn of pure marble, a seeker contemplates the scene. In the distance Mount Fridtjof Nansen rises to an elevation of 13,350 feet above Liv Glacier, flowing from the right rear of the photo. Roald Amundsen's party found passage to the polar plateau on the far side of Fridtjof Nansen, becoming the first to reach the South Pole on December 16, 1911. Eighteen years later, Richard Byrd navigated the first flight to the pole, following a course up Liv Glacier.

Yale University Press books may be purchased in quantity for educational, business, or promotional use. For information, please e-mail sales.press@yale.edu (U.S. office) or sales@yaleup.co.uk (U.K. office).

Designed by Nancy Ovedovitz. Set in Galliard Oldstyle and The Sans Semibold type by BW&A Books, Inc. Printed in China.

Library of Congress Cataloging-in-Publication Data
Stump, Edmund.
The roof at the bottom of the world : discovering the Transantarctic Mountains / Edmund Stump.
    p.   cm.
Includes bibliographical references and index.
ISBN 978-0-300-17197-6 (cloth : alk. paper) 1. Orogeny—Antarctica—Transantarctic Mountains.
2. Geology—Antarctica—Transantarctic Mountains. 3. Transantarctic Mountains (Antarctica)
I. Title.
QE621.5.A6S77 2011
551.43'209989—dc22   2011006403

A catalogue record for this book is available from the British Library.

This paper meets the requirements of ANSI/NISO z39.48–1992 (Permanence of Paper).

10 9 8 7 6 5 4 3 2 1

Publication of *The Roof at the Bottom of the World*
has been made possible by the generous support
of the following benefactors:

Scion Natural History Association
Daryl A. Russell
Bear Gulch Foundation
Harry Rubin, in memory of Morton J. Rubin,
    whose interest in Antarctica took him there
    many times, including fifteen months with the
    Russians at Mirny Station during the IGY
James W. Collinson
Thomas Henderson
Julia and Ralph Maccracken
Julie Smith-David and Scott David
John Splettstoesser
Marjory Spoerri

and
School of Earth and Space Exploration
College of Liberal Arts and Sciences
Arizona State University

This book is dedicated to the memory of my parents,
Sis and Warren,
for encouragement,
for discipline,
and this body that takes me the places I've been.

Contents

Preface

The familiar images of Antarctica include penguins frolicking in their rookeries, endless plains of windswept snow and ice, and tourist or research vessels cruising the straits of the Antarctic Peninsula with its rocky cliffs hung with glaciers and icebergs drifting calmly by. But Antarctica also has a mountain range continental in scale, a remote and desolate wilderness of rock that rises defiantly above the ice. As the Andes are to South America and the Himalayas are to Asia, so are the Transantarctic Mountains to Antarctica. Yet they remain largely unknown, a secret of the few who have had the privilege of journeying there.

The stories of the discovery and exploration of the Transantarctic Mountains span a century of valiant enterprise, from the days of wooden sailing ships and the first sighting by James Clark Ross in 1841, through the heroic era when Scott, Shackleton, and Amundsen vied to be the first to attain the South Pole, to the airborne exploits of Byrd in the 1930s. As expeditions pushed deeper into the interior, the Transantarctic Mountains unfolded before them as a great, linear fortress of rock, dividing the East Antarctic Ice Sheet from the Ross Sea, Ross Ice Shelf, and West Antarctic Ice Sheet. The stories of exploration are familiar to those with polar interest, but the territory, less so.

This book invites you, whether you know the stories or not, to visualize the trails blazed by these explorers who bore first witness to the wonders beyond the icy sea. Following a narrative of the voyages and traverses of those parties that were the first to behold new lands, the figures guide the reader along the routes of discovery that penetrated the unknown at the southern extremity of Earth.

Fundamental to any exploration is the production of maps, as the evidence of where you have been and a guide to those who follow. In this book, I reproduce original maps that charted the succession of discoveries of the Transantarctic Mountains and overlay the actual traverse routes on shaded-relief, topographic maps produced by the U.S. Geological Survey, which show precisely the terrain that the explorers crossed, and, when compared to the original maps, how limited was their view.

During my thirteen expeditions to the Transantarctic Mountains, spanning forty years of research as a field geologist, I have avidly photographed every portion of the mountains that I encountered, with special attention to the explorers' routes, taking some photos from precise locations where those explorers stood. I have culled a representative sampling from my collection of six thousand–plus photographs, most of which appear in print for the first time. I also reproduce select satellite imagery to illustrate the full extent of some of the longest traverses.

Around forty-five million years ago the continent of Antarctica began to pull apart. That portion that we call West Antarctica stretched and thinned and extended away from the stable portion we call East Antarctica. Fifty million years before that event, when Australia and New Zealand were attached to Antarctica, they similarly extended, but then they separated and drifted away, with ocean crust forming in their widening breach. In contrast, when West Antarctica rifted from East Antarctica, it never let go. The crustal thickness of East Antarctica is about twenty-five miles (forty kilometers), typical for continental crust around the world, whereas West Antarctica thinned to around fifteen miles (twenty-five kilometers) during its rifting phase.

The boundary between East and West Antarctica is a fault, a tear along which West Antarctica dropped to elevations below sea level, and the adjacent portion of East Antarctica rose up to form the Transantarctic Mountains. Geologists call such a feature a *rift shoulder,* and the Transantarctic Mountains are the grandest example on Earth. At the time of the upheaval, East Antarctica continued as a plain of low relief from the inland side of the Transantarctic Mountains to its distant shores. In West Antarctica all that remained above sea level were several large islands set some five hundred miles off from the Transantarctic Mountains with an isthmus or archipelago of smaller islands connecting back to the mainland deep in the interior of the continent.

Forty-five million years ago the world was warmer, and Antarctica had no appreciable ice, if any. Soon after the rifting, however, permanent ice began to form on the continent, first perhaps as valley glaciers high in the Transantarctic Mountains. On the East Antarctic side the ice coalesced into an ever-thickening and widening *ice sheet* that expanded to the continental margin, where today it calves into icebergs that disintegrate in the Southern or Antarctic Ocean. In West Antarctica, permanent ice began as a floating *ice shelf.* As it expanded, it thickened and its bottom grounded, forming an ice sheet of its own. Confined between the coastal islands and the Transantarctic Mountains, the West Antarctic Ice Sheet flows in opposite directions from its central high into vast ice shelves that also calve icebergs into the Southern Ocean. The Transantarctic Mountains act as a gigantic dam to the East Antarctic Ice Sheet, which it backs up to elevations of eight thousand feet before spilling over through a series of mighty, *outlet glaciers.* Between northern Victoria Land and Ross Island these glaciers debouch directly into the

Ross Sea, forming a series of floating *ice tongues*. Between Ross Island and Scott Glacier, the outlet glaciers cross the Transantarctic Mountains and add their volume to the Ross Ice Shelf, which maintains a uniform low altitude to a line near the mouth of Scott Glacier, where it is grounded. From there the West Antarctic Ice Sheet gains elevation toward the interior, rising up the face of the Transantarctic Mountains until it reaches a point at the Ohio Range where it merges with the East Antarctic Ice Sheet, inundating the Transantarctic Mountains completely.

Beyond the Ohio Range, the Transantarctic Mountains split, with one branch connecting through the isolated Thiel Mountains to the Pensacola Mountains, which terminate at the Filchner Ice Shelf. The other branch peeks above the West Antarctic Ice Sheet in a string of nunataks (islands of rock surrounded by ice) that include the Whitmore Mountains and Mount Woollard. Most maps of the continent depict the Transantarctic Mountains as including the Pensacola Mountains, whose bedrock geology is similar. The rift shoulder that defines the Transantarctic Mountains, however, follows the line of nunataks into West Antarctica. In this book, I will limit the Transantarctic Mountains to their contiguous portion, which begins at the northern end of northern Victoria Land, spans fifteen hundred miles, and terminates at the Ohio Range.

The topographic maps that trace the explorers' routes deserve special mention. They were produced by the U.S. Geological Survey after the International Geophysical Year (IGY) in 1957–1958, in a rush of activity that charted the entire length of the Transantarctic Mountains at a scale of 1:250,000. Combining aerial photography with ground control, the first printed quadrangles were published in 1959. Topographic engineers who created the initial ground control accompanied overland traverses or utilized helicopter support in coastal areas. Over the next five years, six to eight new maps were added annually to the portfolio of the Transantarctic Mountains. In 1964 the Topographic Division produced an astonishing twenty-four 1:250,000 sheets, covering all the bedrock outcrops of the Transantarctic Mountains, except for five quadrangles, which were published the following year. This bonanza followed a remarkable pair of seasons (1961–1962 and 1962–1963), during which a crew of topographers supported by a pair of army helicopters surveyed the entire fifteen hundred–mile length of the Transantarctic Mountains.

During the period of peak productivity, the personnel at the Topographic Division assigned to the Antarctic quadrangles grew to about 180 participants. The teams split the tasks that included careful measurement and marking of peaks on aerial photos, linking summits to ground control, transferring points to a scribe sheet, measuring and drafting in contour lines, distinguishing ice from rock, adding color, outlining crevasse fields and moraines, airbrushing shaded relief, and inserting geographic names—all the while checking and rechecking everything for accuracy. The contour interval (the elevation difference between contour lines on the map) of these maps is two hundred meters (650 feet). This interval is very coarse when it comes to reckoning on the ground, but cartographers added airbrushed, shaded relief depicting the landscape to a detail that far surpasses that which can be read from the contour lines alone. With burnt sienna bedrock, pale blue ice, and the sun perpetually shining in a northwestern sky, the airbrushed rendering has provided a magnificent base map for tracing the routes of the explorers, showing precisely where they traveled into the mountains, through crevasse fields, around ridges, over passes, up outlet glaciers, and onto the polar plateau.

Yet for all the detail on the 1:250,000 quadrangles, their makers could not have imagined the accuracy that would come to cartography in the fifty years after the International Geophysical Year. Today, arrays of satellites allow us to determine to within millimeters our position on the globe. Other satellites provide images of the surface of the Earth in both visible and invisible wavelengths, giving us a view of the surface of the Earth in incredible detail, now available to the public as readily as clicking Google Earth. A seamless Landsat view of Antarctica has recently been released (http://lima.usgs.gov/). This is the source of the imagery in the figures of the Transantarctic Mountains and the continent as a whole that follow the Acknowledgments. The Landsat satellite maintains a nearly polar orbit, so that as it orbits the globe, it passes over a series of tracks that all but encircle the planet. Because the orbit is slightly skewed from the poles, the coverage in Antarctica reaches only up to 82° 30′ South. Beyond that, the interior of Antarctica is filled by less detailed imagery from a MODIS satellite (Moderate-resolution Imaging Spectroradiometer), and remains the most poorly imaged region on Earth.

So whether you have your computer with its Landsat imagery at hand, or allow only this book to be your guide, step back in time, to a time before satellites, to the era when the last, blank space of white on maps encircled the southern extremity of the globe and when our encounters with nature were much more direct.

Acknowledgments

Typically, the first time one goes to the Antarctic for research, it is with someone else's party. I still vividly remember the phone call in April of 1970, from Ohio State University, inviting me to interview for a position on a large, helicopter-supported project to the Transantarctic Mountains. The voice with a British accent at the other end of the phone belonged to David Elliot, at that time a young postdoc at the Institute of Polar Studies. The fieldwork begun that season would become the basis for my dissertation and for my academic career. David was the person who opened the door for me to the polar wilderness that has been my passion for the past forty years. Thanks, David, from the bottom of my heart.

All research that is conducted through the U.S. Antarctic Program (USAP) is funded by the National Science Foundation, specifically by the Office of Polar Programs. Since NSF is a federal agency, I must also thank the U.S. taxpayer for the opportunity to do research in Antarctica. My own research has been based primarily in the field, mapping and collecting rocks that are analyzed upon return. My photo collection is replete with boring shots of outcrops recording collecting localities and geological relationships, but I have always also taken photos of the scenery that passes in the course of my work. As with any project funded by NSF, mine were judged worthy by peer review, and the payback was publication in the scientific literature. But these results seldom reach the public. Through this book I hope to share the magnificence of this land.

With great pleasure I acknowledge Paul Dalrymple, the irascible bard of *The Antarctican Society Newsletter* and one of the original winter-overs during the IGY. Paul was in the audience some twenty-five years ago when I gave the annual lecture to The Antarctican Society in Washington on the subject of exploration of the Queen Maud Mountains. Ever since then he has bugged me, sometimes openly in the newsletter, "When are you going to write that book?!" Well, Paul, here it is.

John Splettstoesser, another OAE (Old Antarctic Explorer), has been a great supporter throughout my career, from the time I was a fresh grad student at Ohio State and he was the assistant director of the Institute of Polar Studies, to recent years, when he served as one of the reviewers on this book. John, sincere thanks for all the times you have been there.

Over the years I have greatly enjoyed my interaction with a number of members of the Topographic Division of the U.S. Geological Survey, who have been involved in one way or another in the production of Antarctic maps. One who deserves special mention is Robert Allen. Bob was part of the group that produced the 1:250,000 quadrangles in the late 1950s and early 1960s. He introduced me to the aerial photography collection, and every time that I have gone back to Reston to use the browse files, I have enjoyed his chatty, seemingly endless knowledge of Antarctic mapping trivia.

In the course of writing this book, I also was honored to have interviewed both the late Bill Chapman and Pete Bermel, who had led the Topo North and South and the Topo East and West surveys, respectively, in the early 1960s. Both were gracious and unassuming in sharing their highflying stories with me.

My personal heroes in Antarctic exploration are the three members of Blackburn's party that traversed to the head of Scott Glacier. This led me to the National Archives, where Quin Blackburn's field notes and journals are kept. It was with keen interest that I was able to trace the party's route and read his observations and thoughts about the geology and the terrain that they discovered. I had finished Chapter 6 based mainly on Blackburn's journals but felt that it needed fleshing out with more personal material. In late 2007 the publication of Stuart Paine's diaries from the Second Byrd Antarctic Expedition, edited by his daughter M. L. Paine, came as a godsend. This book offered a detailed, personal counterpoint to Blackburn's geological notes, and is interesting for its perspective on life at Little America during the winter-over when Byrd occupied and then was rescued from the advanced weather station on the Ross Ice Shelf.

The satellite imagery for all figures except those in the Introduction was taken from the gallery of images provided by NASA's MODIS Rapid Response System (http://rapidfire.sci.gsfc.nasa.gov/). This resource produces, sometimes on a daily basis, imagery of such phenomena as global wildfires, hurricanes, dust storms, plankton blooms, and the disintegration of Antarctic ice shelves. The figures of the continent that follow were adapted from the Landsat Image Mosaic of Antarctica (LIMA) (http://lima.usgs.gov/).

That this book exists at all is due to my agent, Regina Ryan. After her first read, she said she couldn't see what I was trying to say, and that if it was to work I needed to have better maps. This led me to the shaded-relief maps and a year of scribing little, curvy lines with Adobe Illustrator. Her hand expertly crafted the proposals. She further nudged me into writing the set of sidebars, then took a scalpel to them. From start to finish she has been a great ally and advocate, and a trusted keeper of the faith.

That this book has surpassed my hopes and expectations is due to my editor, Jean E. Thomson Black. She enthusiastically embraced the concept from the start, and successfully maneuvered the project with full color up the line at Yale University Press. Her deft editing has turned the stodgy prose of a longtime writer of science articles into a narrative that flows. In the autumn of 2010, my return to Antarctica imposed a stringent deadline for bringing together the edited figures and manuscript. Jean and her staff worked above and beyond the call to push it through to production. The extent to which this book is a success is due largely to her efforts.

My first season in Antarctica was 1970–1971. I met my partner for life, Harriet Mac-cracken, in the run-up to my second season, 1974–1975. Since then she has steadfastly kept the home fires burning during my many absences. When we started having children in the mid-1980s, she traded the solitude of our separations for the heroically hard work of a single mom. As they grew older, Simon, Molly, and Nick came to understand and share my passion for that frozen land at the end of the Earth. By the 2010–2011 season, the last of the kids has gone off to college, and we have come full circle. It is with love and deep gratitude that I acknowledge my family's support through all the years. Simon, Molly, Nick, you are the joy in my heart. Harriet, you are my rock.

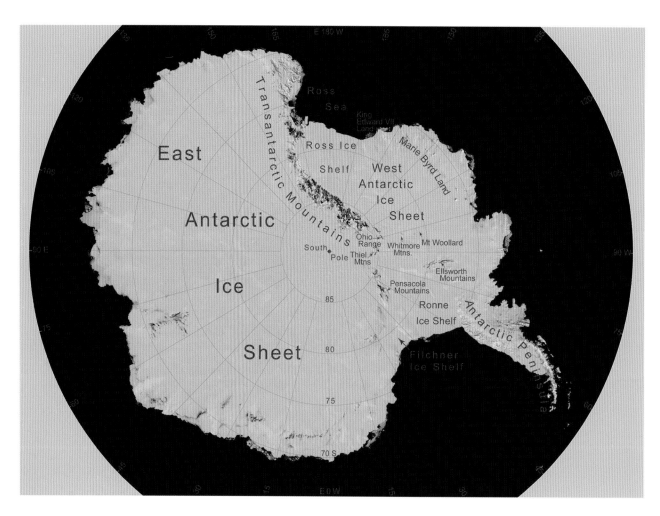

Satellite image of the continent
of Antarctica

(opposite) Satellite image of the
Transantarctic Mountains

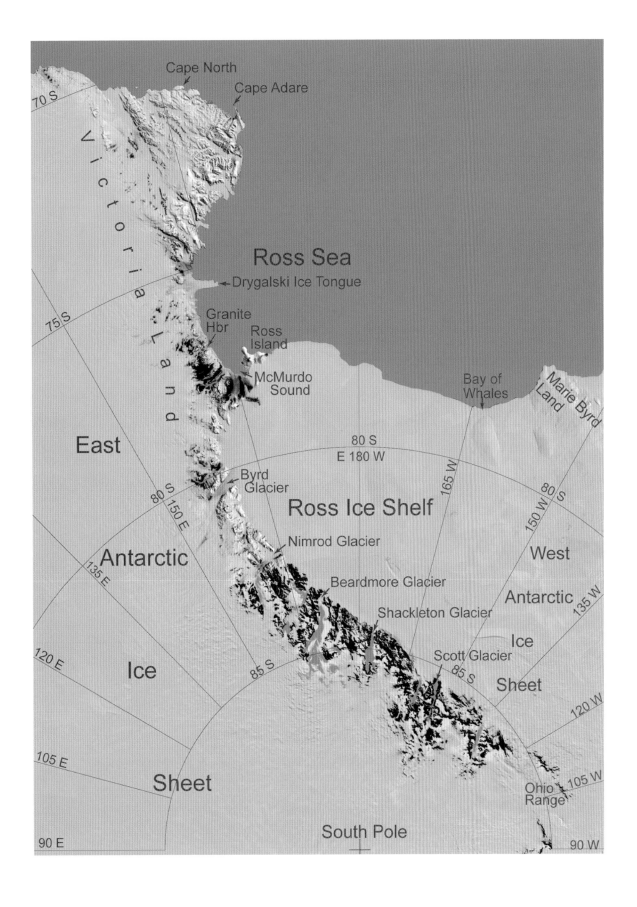

I Through the Portal
Discoveries along Coastal Victoria Land

In the aftermath of the Renaissance and three centuries of unprecedented exploration by Europeans, the world had become a sphere encircled by an ocean, and its edges had vanished. Previously, blank spaces beyond the boundaries of maps were limitless and often depicted with demons or dragons. Now the world was finite, and the blank spaces on its face were steadily fleshed out as explorers circled the shorelines of continents, probed deep into their river systems, and happened onto islands as they crossed the open seas. By the end of the eighteenth century, much of the Earth was charted. Certainly substantial portions of interior Africa and Australia were still unknown, but at least their continental outlines had been drawn. The lone remaining large field of white on the globe lay above 60° S. Conceived as a balance to the northern ocean, speculations of a southern continent, Terra Australis Incognita, reached back to Classical times. When Magellan sailed through his straits in 1520, some cartographers interpreted the southern wall of the channel to be a promontory of that continent, and they mapped it accordingly (Fig. 1.1). Fifty years later, when Drake sailed around Cape Horn, proving that the "promontory," Tierra del Fuego, was no more than an island at the tip of South America, what lay beyond the Southern Ocean remained a mystery.

During the second of his celebrated voyages (1772–1775), James Cook encircled the Antarctic, assaulting it in all quadrants, only to be rebuffed by icebergs and pack ice. His deepest penetration was to 71° 10′ S at longitude 106° 54′ W. Because icebergs generally originate on land, Cook was of the opinion that a continent existed beyond the ice edge, but for all he really knew the ice at the South Pole covered an ocean, as it did in the Arctic.

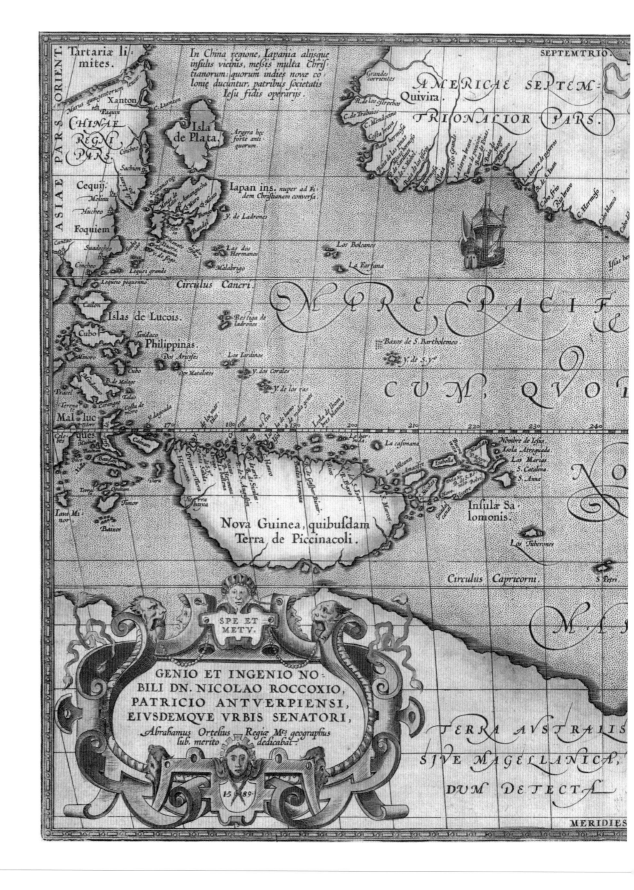

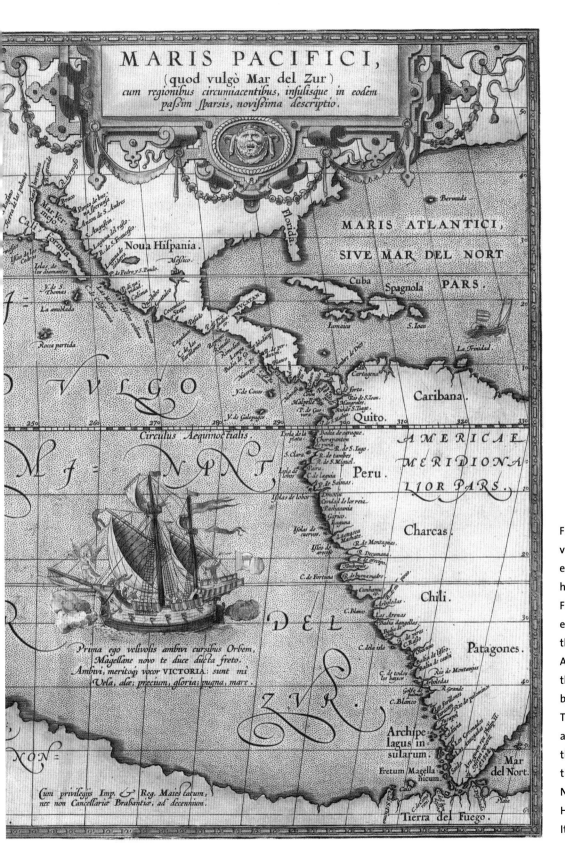

Figure 1.1. After Magellan's voyage around the southern end of South America, his sighting of Tierra del Fuego to the south was extended to encompass the whole of a supposed Antarctic continent. In this map published in 1589 by Abrahamus Ortelius, Terra Australis encircles and reaches northward to the Tropic of Capricorn in the area to the south of New Guinea. Published by Historic Urban Plans, Inc., Ithaca, New York.

The maps drawn after Cook's voyages showed an "Antarctic Icy Sea" encircling the globe to the north of 70° S, but from there to the South Pole they charted only emptiness.

During the early nineteenth century, Antarctica continued to lure explorers to its icy fringes. The motivations of those who sailed south varied, but they all shared the risks that awaited them. In a passage that applied to the explorers of 1830 as well as it did to those of 1930, Laurence Gould concluded in answering the question of why the members of Byrd's first expedition had come south,

> Some came for the sheer love of adventure and wanted no reward beyond that; some wanted fame or its counterfeit, publicity; some were mercenary and thought primarily in terms of what they were going to get out of it; and lastly there was that small group, the like of which gives character to any expedition of merit—not necessarily scientists at all, but men who could understand the lure, if not the love, of knowledge for its own sake; men who came not for position or money but who found full reward for their effort in the pursuit of an ideal.

Thaddeus von Bellingshausen led a Russian national expedition of discovery in 1819–1821. Taking an approach similar to Cook's, he circumnavigated the ice margin, sailing into it where conditions were favorable. On January 16, 1820, with minimal visibility because of a snow storm, he reported a solid stretch of ice (probably an ice shelf) at the coastline of present day Dronning Maud Land (a few degrees west of the Prime Meridian) (Fig. 1.2). On February 5 he made a clearer sighting of fast ice with high vertical cliffs 20° to the east. These were the first marks to be placed on a map at the edge of East Ant-

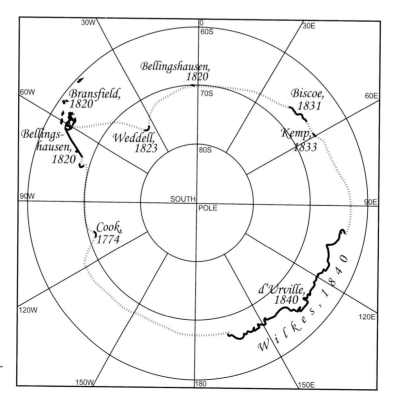

Figure 1.2. By the austral summer of 1840–1841 the outline of the Antarctic continent was beginning to take shape. The broadest margin that remained unknown was from Alexander Island to 165° E.

arctica. The significance of this discovery, however, was not widely recognized outside of Russia until the twentieth century, when Bellingshausen's report was translated into German (1902) and then English (1945).

During the 1820–1821 season, Bellingshausen sailed east from Sydney, Australia, and discovered "Alexander I Land," now known to be an island (Alexander Island), connected to the foot of the Antarctic Peninsula by fast sea ice. From there he sailed north to the South Shetland Islands, where he encountered American and British sealing ships that had rushed to the islands after their discovery in 1819 by William Smith. Chasing favorable winds, Smith, captain of a cargo ship rounding Cape Horn, had sailed farther south than anyone before and happened onto this rich breeding ground of pinnipeds.

The year after Smith's discovery, the British Admiralty sent a surveying expedition commanded by Edward Bransfield with Smith as guide. On January 30, 1820, they had charted the rocky peaks of part of the northern end of the Antarctic Peninsula (Trinity Peninsula). The coincidence in time and the sighting of ice versus rock have led to long-standing arguments whether Bellingshausen or Bransfield was the first to sight the continent of Antarctica.

Although the early sealers and whalers who probed first the Antarctic Peninsula and later more distant coasts sailed foremost for commercial enterprise, many of the captains also had a keen interest in natural history and in the discovery of land for its own sake. The British whaling and sealing firm of the Enderby Brothers particularly encouraged this sort of activity. One of their captains was James Weddell, who sailed south in search of seals to the east of the Antarctic Peninsula, penetrating deep into the sea that would come to bear his name. On February 20, 1823, he reached 74° 15′ S and turned and headed north, claiming the record of "farthest south" that would stand until 1841, when Ross bested it. Although Weddell sighted no land, he had mapped ocean to a considerably higher latitude than the coastline spotted by Bellingshausen.

John Biscoe, another Enderby Brothers captain, sailed close along the ice-bound coast of East Antarctica previously sighted by Bellingshausen, and on February 28, 1831, discovered a coastline with rock jutting through the ice at 66° S, 51° E, which he named Enderby Land.

In the late 1830s France, the United States, and Great Britain mounted national expeditions, each charged to explore high southern latitudes. The French voyage, captained by Jules Sébastien César Dumont d'Urville, sailed south from Hobart in the summer of 1839–1840 and sighted land at 142° E, 67° S on January 19, 1840. A party landed on an islet, collected granite and several startled penguins, claimed the new land for France, and toasted with a bottle of Bordeaux. The expedition then sailed west as far as 134° E before returning to Hobart. Dumont d'Urville named this new territory Adélie Land after his wife, and subsequently the penguins that the expedition collected were named Adélies.

The United States Exploring Expedition of 1838–1842 commanded by Charles Wilkes was a complicated voyage involving a fleet of six ships that separated and came back together, visited all seven continents, and encircled the globe. One ship returned ill-suited for the high seas; another went down in a storm. Wilkes had been ordered to sail south from Hobart in the summer of 1839–1840 until he encountered coastline and then to explore westward as far as 45° E. Three of the ships reported sighting land on January

16, 1840, at approximately 164° 30′ E, 64° 11′ S, but none recorded it in the log until January 19, thus setting up an acrimonious set of claims and counterclaims between Wilkes and Dumont d'Urville regarding who had seen and done what first. This argument was exacerbated by an earlier incident when *Porpoise,* one of the Exploring Expedition's ships, passed *Astrolabe,* Dumont d'Urville's ship, in a fog and neither one trimmed sails to slow down for the other. Wilkes managed to stay his course close to the coast, sighting land ice at a number of locations before reaching a large ice tongue (Shackleton Ice Tongue) at about 100° E. From there he sailed back to Hobart, having followed the coast for seventeen hundred miles.

By the end of the austral summer of 1839–1840 the outline of Antarctica had taken on much of the shape that we know today (see Fig. 1.2). The bow-shaped coastline of East Antarctica was known to lie close to the Antarctic Circle from 164° E to the Prime Meridian, thanks to the voyages of Bellingshausen, Briscoe, Kemp, Dumont d'Urville, and Wilkes. Most sightings were of fast ice, but in the few places that outcrops of rocks were seen, all were of low elevation and low relief. In contrast, towering peaks rising out of ice-choked straits characterized the distal end of the Antarctic Peninsula projecting toward South America at approximately the 60° W meridian. To the east of the peninsula, Weddell had shown that the ocean reached as far south as 74° S, considerably farther than the coastline to the east of the Prime. The longest remaining segment of unknown coast was from Alexander Island to 165° E, Wilkes's starting point. Cook's farthest south had been along longitude 106° W, to 71° 10′ S, but floating ice was all that anyone had recorded in that quadrant of Antarctica.

Such was the map of the southern continent as one of the greatest players in Antarctic exploration was about to enter the game. With the endorsement of the British Association for the Advancement of Science, the government organized an expedition for the study of magnetism at high southerly latitudes and the exploration of new lands in the Antarctic. The command could not have been given to a more capable captain than James Clark Ross, a veteran of Arctic voyages for twenty years. Skilled in operating magnetic instrumentation, Ross had been the first person to reach the Magnetic North Pole. He joined the Royal Navy at the age of eleven and rose through the ranks, mentored both by his uncle John Ross, who was involved in expeditions in search of the Northwest Passage, and by Edward Parry, another giant of Arctic exploration.

In the early 1830s the Rosses were stuck in Prince Regent Inlet in northern Canada, beached on the Boothia Peninsula, with their ship *Victory* in the ice just offshore. By the end of their saga, they had wintered over four years, and they were rescued on the high seas by a chance encounter with a whaling ship, after dragging their small life boats over hundreds of miles of rough sea ice.

In the spring of 1831, between the second and third winters (when their situation was grim but not yet desperate), James Clark Ross had set out with an Eskimo sledging party to explore the region, having learned Inuit during his expedition with Parry. Equipped with the instrument (a clinometer) to measure the dip of magnetic lines, Ross worked his way along the western shore of the Boothia Peninsula, and on June 1, 1831, measured a magnetic inclination of exactly 90°, thereby locating the Magnetic North Pole.

If the Antarctic pack ice could be penetrated, the prize for Ross would be also to reach the Magnetic South Pole. Age forty, he was in his prime and said to be the most handsome officer in the Royal Navy. As the expedition was being organized, Ross had stated his intention of sailing south along 160° E. When he arrived in Hobart in August 1840, he was dismayed to find that he had been preempted the previous season by both Dumont d'Urville and Wilkes, who in that time had made important discoveries along a sizable segment of coastline. Bound not to follow in the paths of these less experienced rivals, the irascible Ross made the fateful decision to sail south along 170° E in search of new lands and a passage to the Magnetic South Pole.

The British expedition sailed with two well-outfitted ships, HMS *Erebus* and *Terror*, both bomb vessels (370 tons and 340 tons, respectively) with their decks stoutly fitted for the recoil of mortar cannons and their hulls reinforced several years earlier for ice rescue in the Arctic. The ships were a bit slow but would be tough and reliable when it came to the pounding. They sailed from Hobart on November 12, 1840, spent three weeks in the Auckland Islands, and then continued south. On the last day of 1840 the ships had their first encounter with the pack, to Ross's view not as formidable as he had been led to believe by the reports of others.

Pack ice is composed of a collage of broken blocks of sea ice born of the previous winter's freeze (Fig. 1.3). By spring, Antarctica is surrounded by a frozen shell of ice floating on the Southern Ocean, which begins to break up along long cracks, or leads, that penetrate into its interior. As the breakup continues, pieces of ice bump and grind against each other, blunting their corners and fragmenting into smaller polygons. Storm swells do the most destruction. The ice floes are six, maybe ten feet thick, and tens of feet to miles across. Driven in part by currents and in part by the wind, the pack reacts animately, contracting from wind pressure at its edges and expanding when the air is calm. The pack also dampens swells, so to be deep within it may be haven from a storm. But closer to the margins when winds are severe, hefty waves may roll through with a great gnashing of ice.

When *Erebus* and *Terror* first encountered the pack, a freshening gale blew the ships back north into a lane of icebergs that they had previously crossed. There they rode out a spell of snow squalls under easy sail, dodging bergs in moments of high anxiety, but by January 3, 1841, the storms had passed, and Ross was back at the edge of the pack. From the crow's nest, the view was that of a vast continuum of ice, merging into the horizon in both southerly quadrants with not a blip to disturb the distant surface. The margin of the pack was composed of small, tightly fitted floes, but beyond that the lookout could see black, edgy lines between the white blocks that branched and merged deeper into the pack (Fig. 1.4). Somewhere in this maze *Erebus* and *Terror* would have to force a passage to the interior, nudging and jostling their way between the shifting floes. These were wooden sailing ships, before the widespread use of steam power, which gave a ship the ability to reverse its propeller and back out of a tight situation. There would be no turning back.

Ross sailed along the margin for several miles, finding no one spot better than the next, so he boldly gave the signal for *Terror* to follow as *Erebus* turned straight into the pack and "bore away before the wind." With a crack like rifle fire and a lurch, the ship hit the pack, gave only a little shiver, and plunged forward, the mate in the crow's nest

Figure 1.3. (opposite) Drifting off the coast of northern Victoria Land, loose pack ice offers less of a challenge to ships seeking passage to the south than tight pack or narrow leads. Backlit by the sun are the three states of water—vapor, liquid, and ice.

Figure 1.4. When Ross approached the edge of the ice pack on January 3, 1841, he found it tightly packed with small floes along its edge. Heading straight into the pack, *Erebus* and *Terror* ground through the resistant perimeter, then broke into more open water.

signaling directions. The ice was harsh, but the thumps and scrapes reached a sort of cadence, and the ships moved through.

For the next four days *Erebus* and *Terror* alternately drifted with a tight pack during snow squalls or made slow progress under light winds. At one calm period Ross descended onto an ice floe and took magnetic measurements, 68° 28′ S, 176° 31′ E, with a dip of 83°, putting the South Magnetic Pole 500 miles to the southwest. Early on January 9 the ships quite suddenly broke out into open water, having traversed 134 miles of pack. With a freshening gale they ran with the wind for almost 30 miles, then trimmed sails and hove to. By noon the next day the storm had retreated, and with high hopes Ross pointed his ships directly toward the South Magnetic Pole.

But no sooner had the ships gotten under way than the watch reported a land blink, a phenomenon that occurs when land beyond the horizon radiates reflected sunlight, producing a narrow, bright band at the horizon, a perceptible aura of mountains unseen. At 2:30 A.M. on January 11 they spotted land during the middle watch—a thin escarpment beyond the dark water, perhaps 100 miles away, that shimmered and then was lost in the haze. By 9:00 A.M. the fog had lifted, and Robert McCormick, the surgeon and natural historian on the *Erebus*, was in the crow's nest sketching the scene as it materialized beneath an azure sky. He wrote, "The weather was all that could be desired for giving effect to such a magnificent panorama, as gradually unfolded itself like a dissolving view to our astonished eyes." A vast mountain range struck off to the southwest then arched back to the southeast. Crowning the range was a line of summits completely mantled in snow, except for thin, dark lines at some of the peaks and ridgelines. The lower reaches of the mountains formed a cliff of volcanic-appearing black rock that was festooned with horizontal streaks of snow. Around the northern tip of the cliff there appeared to be a bay that would offer the ships safe harbor, but a heavy swell broke on a tight pack blocking the entrance. From there they steered to the southeast, rounding the black cliff and sailing south along the coast for several hours before lying farther offshore for the night (Fig. 1.5).

Ross recorded the sights:

It was a beautifully clear evening, and we had a most enchanting view of the two magnificent ranges of mountains, whose lofty peaks, perfectly covered with eternal snow, rose to elevations varying from seven to ten thousand feet above the level of the ocean. The glaciers that filled their intervening valleys, and which descended from near the mountain summits, projected in many places several miles into the sea, and terminated in lofty perpendicular cliffs. In a few places the rocks broke through their icy covering, by which alone we could be assured that land formed the nucleus of this, to appearance, enormous iceberg.

Ross bestowed the name Admiralty Range on these lofty mountains. The apparently highest summit he named Mount Sabine, after Edward Sabine, foreign secretary of the Royal Society and an ardent supporter of the expedition. Because of its closer proximity to the shore, Mount Sabine at 12,200 feet was perceived to be higher than a pair of peaks in the interior of the Admiralty Range. Mount Minto (13,660 feet) and Mount Adam (13,153 feet), named for high lord commissioners of the Admiralty under whom Ross served, are actually the highest summits in all of Victoria Land (Fig. 1.6). Ross named the cape at the northern end of the dark cliff Adare, after Viscount Adare, a friend of Ross, and the bay behind it Robertson Bay, after the surgeon on the *Terror*.

The South Magnetic Pole lay about five hundred miles to the southwest beyond the mountains. There at Cape Adare, at the corner of this new land, the choice was to sail either west or south. Which direction might better offer a waterway to the pole?

The next day while sailing close to shore, the ships were caught in a strong tidal current. To escape, they slipped into the lee of a small island. A narrow boulder beach offered possible access, so two boats were let down with a landing party. Captain Ross was the first out of the bow, with the others quickly following. With the simplest of ceremonies, the party planted a flag and Ross claimed the island for Mother England. Liquor was passed around and everyone drank a toast to Her Majesty, the young Queen Victoria, and another to Prince Albert. At 71° 56′ S, 171° 07′ E, it remains to this day "Possession Island." The island was composed of a dark volcanic rock called basalt and was home to several tens of thousands of cheeky Adélie penguins that pecked at the men's legs as they made their way between the birds' rocky nests.

For the next two weeks, the expedition worked its way south, trying to stay in close to the mountains, but gales and ice floes repeatedly forced them away. On good days, when the weather cleared, the ships would work as close as they dared, nosing toward indentations in the coastline, but each was filled with an apron of fast ice that rimmed the shore. On one of these days when they were able to see the Admiralty Range with clarity, Ross wrote, "we gazed with feelings of indescribable delight upon a scene of grandeur and magnificence far beyond anything we had before seen or could have conceived"

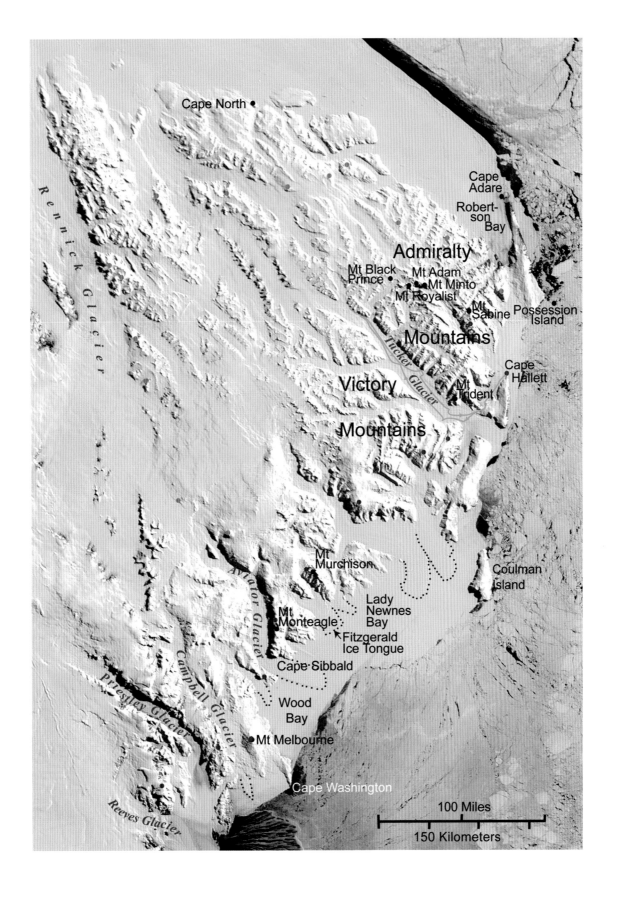

Cape North •

Cape
Adare

Robert-
son
Bay

Admiralty

Mt Black   Mt Adam
Prince • • • Mt Minto
      Mt Royalist
              Mt
            • Sabine   Possession
                          Island

Mountains

Tucker Glacier

Rennick Glacier

Victory

Mt
Trident

Cape
Hallett

Mountains

Mt
Murchison

Coulman
Island

Aviator Glacier

Lady
Newnes
Bay

Mt
Monteagle

Fitzgerald
Ice Tongue

Cape Sibbald

Campbell Glacier

Priestley Glacier

Wood
Bay

Mt Melbourne

Cape Washington

Reeves Glacier

100 Miles

150 Kilometers

Figure 1.6. This aerial image of a portion of the Admiralty Mountains is viewed to the east in the opposite direction from the sightings by Ross and later explorers. Mount Minto, the highest point in northern Victoria Land, is the dark, faceted peak in the upper right of the image. Mount Adam is the prominent peak a short distance to the left at the culmination of three ridgelines. The dark massif with the rounded summit directly in front of Mount Adam is Mount Royalist. The dramatic peak halfway along the ridgeline to the left is Mount Black Prince. Mount Royalist and Mount Black Prince were first sighted in 1958 by the New Zealand geological party working on Tucker Glacier (compare Fig. 1.5). The Adare Peninsula is the dark cliff that subtly rises and falls on the far side of Robertson Bay in the rear of the image. This is the long, volcanic ridge noted by McCormick as the ships approached the land. Robertson Bay was full of tight pack, which prevented their entering. Ridley Beach, the site of Borchgrevink's winter-over, is at the northern (left) tip of the peninsula.

(Fig. 1.7). More often than not, the peaks would fade in and out of view as clouds overtook them, and the ships would give more distance to the land. One high, conical peak well to the south, glimpsed on several occasions, reminded the men of Mount Etna, the famous volcano on the island of Sicily.

On January 22 the ships crossed 74° 15′ S, surpassing Weddell's record for southernmost travel. Ross wrote, "An extra allowance of grog was issued to our very deserving

Figure 1.7. On a clear day Ross viewed this portion of the coastline of northern Victoria Land, describing it as "a scene of grandeur and magnificence." The prominent cliff face slightly to the left of center is Cape Sibbald, and the prominent peak near the right margin of the image is Mount Monteagle. The Aviator Ice Tongue issues into the frozen surface of the Ross Sea from the valley behind Cape Sibbald. The peak near the coastline to the left of the Aviator Ice Tongue is the site where, on New Year's Day 1962, the Topo North survey party was nearly blown off its station by katabatic winds (Chapter 7).

crew." It being Saturday, they joined in the seaman's toast to "Sweethearts and Wives— May the sweethearts become wives and the wives remain sweethearts." Then another gale struck and drove them off to the east for the next two days. Returning from the west the ships encountered tight pack ice forty or fifty miles out from the shore. The crew had clear views of the massive volcanic cone rising almost nine thousand feet out of the sea, which was named Mount Melbourne, after Lord Melbourne, the British prime minister at the time.

At noon on January 26 the magnetic field dipped at 88° 33′, putting the pole only 174 miles west of south, but Mount Melbourne lay only 40 miles to the west, and although south of it the coast drew back to the west, farther on it straightened and headed due south along the meridian. Rather than try to penetrate the pack that was holding them off from the mountains, Ross ordered his ships south, with hope of still rounding the range and coming up on the magnetic pole from the south. But soon the expedition was putting distance between itself and the pole. The next day Ross made a rough landfall at Franklin Island, another small volcano of basalt, and again claimed possession. Thence the ships headed south toward a peak that had appeared that morning and loomed increasingly large across their course.

Ross recorded, "'High Island'; it proved to be a mountain twelve thousand four hundred feet of elevation above the level of the sea, emitting smoke and flame in great profusion; at first the smoke appeared like snow drift, but as we drew nearer, its true character became manifest" (Fig. 1.8). How wondrous it must have been to discover this fountain of fire in a land of ice! The active volcano was joined to an inactive one, rising to 10,900 feet. Ross named the twin cones Mount Erebus and Mount Terror, after the expedition's ships. To Ross they appeared to be connected to the mountain range to the west, a minor error in an exceptional charting of new lands. With the south now blocked, the only possible avenue to the pole would have to be found to the east, but as they sailed farther south, an even more phenomenal specter rose out of the sea:

Figure 1.8. When Ross discovered Mount Erebus, the world's southernmost active volcano, it was "emitting smoke and flame in great profusion." The volcano has remained active to the present day, with its plume of vapor wafting from the summit.

As we approached the land under all studding sails, we perceived a low white line extending from its eastern extreme point as far as the eye could see eastward. It presented an extraordinary appearance, gradually increasing in height to a perpendicular cliff of ice, between one hundred fifty and two hundred feet above the level of the sea, perfectly flat and level at the top, and without any fissures or promontories on its even seaward face. What was beyond that we could not imagine; for being much higher than our masthead, we could not see anything.

As the ships closed on this wall of ice, Ross sighted what appeared to be high summits to the south of Mount Terror, but where they went was not observable. In a gesture of symmetrical favor, he named them the Parry Mountains after his mentor, who had bestowed Ross's name on the northernmost known land some years before.

Ross's expedition tracked "the Barrier" more than three hundred miles to the east. With soundings up to 410 fathoms (2,460 feet), Ross concluded correctly that the front, at least, of this gigantic ice wall floated and was not based on land. After inventorying the effects of temperature changes and storms on the breakup of thick ice in the Arctic, he wrote, "But this extraordinary barrier of ice, of probably more than a thousand

THE TRIP DOWN

Having taken seventy-eight days from the time he departed Hobart, Tasmania, until reaching Ross Island, James Clark Ross could not have imagined the ease with which scientists of the modern era come to Antarctica. In October 1970 I boarded an air force C-141 Starlifter in Christchurch, New Zealand, and six hours later was standing on the ice runway at McMurdo Station, the doorstep to the interior of Antarctica. The Starlifter was a big jet transport plane with swept wings that drooped from the top of its fuselage. From October to mid-December, these wheeled aircraft can land on the smooth, hard sea ice out from McMurdo Station. By January, this seasonal ice typically is breaking out into open water, and air operations move several miles south onto the Ross Ice Shelf, where only ski-fitted planes like the C-130 Hercules can land on the softer, snow runway.

McMurdo Station is the hub of U.S. operations on the continent, built around the bowl of Winter Quarters Bay, where Robert Falcon Scott was based in 1902–1904. It receives and sends out a fleet of ships, planes, and helicopters, all in support of projects funded by the National Science Foundation. Hercules aircraft routinely fly to the South Pole, and, during some seasons, supply remote helicopter camps more than five hundred miles from McMurdo (Fig. S.1).

The day I took off for Antarctica, the center of the Starlifter was filled with a series of pallets laid full of high-priority cargo. The passengers, a mix of scientists and navy personnel, were seated on two rows of sling seats facing the cargo, with barely enough room for one's knees. We crammed into our seats, dressed

Figure S.1. The remote, helicopter-supported camp deployed in northern Victoria Land during the 1981–1982 field season. The six structures on the left side of the camp are Jameswaysns, or portable canvas Quonset-like huts used as berthing facilities, a mess hall, a science laboratory, and a generator shack. The cargo yard extends to the right from these. The disrupted snow at the extreme right is a snow pit dug to supply the snow melter for the camp's water supply. An LC-130 Hercules aircraft sits next to the camp runway. The two shiny objects immediately above the Herc are fuel bladders, giant "water beds" filled with jet fuel for the helicopters. Two of the three HU-1D helos assigned to the camp sit on the pad. I am aboard the third as it approaches camp.

in full survival gear, including parka, windpants and liner, "bear-paw" mittens, and the white, rubber footwear that everyone calls "Bunny boots," because they resemble Bugs Bunny's feet.

The crew chief, a fifty-something sergeant with a potbelly and a one-piece jumpsuit zipped open way too far, briefed us on safety procedures. As the plane's engines began to whine, I twisted in my earplugs, retreating to a private world of adrenaline. I couldn't believe my good fortune in having been selected to go to Antarctica, yet here I was buckled into a plane about to take off across the Southern Ocean and to land somewhere on the other side at an outpost at the end of the Earth.

At the back of the Starlifter was a narrow floor space between the cargo ramp and the end of the last pallet where passengers could stand and stretch cramped muscles. A pair of cargo doors flanked this space, each with a single, circular window about a foot and a half in diameter, which offered those of us not on the flight deck our only glimpse of the exterior. Several hours into the flight I went back to look out the windows. The ocean was dark blue spotted with white blocks of ice that soon coalesced into a solid sheet of white. I stood rapt at the window as land appeared, several low ridges rising above the sea of ice. Ross had sailed along this coastline in 1841 late in the season with all of the sea ice broken out.

The flight path from Christchurch to McMurdo crosses northern Victoria Land approximately along the 167° E meridian. Since no one else seemed particularly interested, I jumped back and forth between windows, scanning the vast network of ridges in the central part of the region. Out the right side of the plane I could see the backsides of the Admiralty and Victory Mountains that had stood as a high, peaked wall to Ross. At about Lady Newnes Bay, the Starlifter reached the coastline with its towering peaks, steep glaciers and ice tongues protruding out into the Ross Sea. I stayed glued to the window on the right side of the plane. Mount Melbourne came into sight, then the spectacular Drygalski Ice Tongue, not discovered until 1902 by Scott. And then we were told to take our seats in preparation for landing.

The plane landed smoothly by Antarctic standards, and soon I was stepping down onto the ice of Antarctica. My first impression was of brilliant, white light. After riding in the dimly lit cargo compartment for so many hours, I emerged into a day of hazy white sky and intense sunlight. I remember not being able to see anything beyond the various buildings, planes, and tracked vehicles that populated the Ice Runway a couple miles out from McMurdo. My senses were bombarded by engine sounds from all directions, and the acrid, oily smell of jet fuel. This wasn't the Antarctica I had anticipated, but rude civilization. This noisy, smelly outpost was simply the gateway to a wilderness that would be beyond my wildest dreams.

feet in thickness, crushes the undulations of the waves, and disregards their violence: it is a mighty and wonderful object, far beyond any thing we could have thought or conceived." As the season drew to an end, snow squalls and heavy pack were joined by freshly forming sea ice. On February 9, just as Ross was deciding to turn back, he saw a small bay opening in the Barrier to the south, and so struck for it. At this low point the men could see through for the first time onto the flat, featureless expanse behind the wall. Fresh, pancake-shaped floes of ice were freezing all around the ships, so the expedition retreated.

The ships retraced their path to Franklin Island and proceeded westward in an attempt to reach land. The view to the south was one of continuous high ground connecting Mount Erebus with the mountains to the west. The expedition named this broad reentrant in the ice front McMurdo Bay, after the senior lieutenant on the *Terror*. The ships were able to push west to within ten or twelve miles of the coast, but a tight pack stopped their forward progress. Off to the northwest a point of land suggested a possible winter haven. Two hours of hard pushing closed no distance on this cape, so after consulting with Commander Francis Crozier, captain of the *Terror*, Ross decided to abandon the quest. They had found no safe harbors anywhere along coastal Victoria Land from which a land assault could be staged the following spring. They made a final landing attempt on February 21 at Cape Adare, but a solid pack held them off about eight miles, so they set a course to the northwest along the further coast. For more than one hundred miles they charted a series of capes with relatively low mountains behind them. The westernmost promontory shielded a small bay full of small icebergs (Fig. 1.9). Ross named it Cape North, because it appeared that beyond it the coast dropped back to the south of

Figure 1.9. Icebergs that calved the previous season clog the ice-bound bay at Cape North. This was the westernmost point of land sighted by Ross on the return leg of his voyage of discovery in 1840–1841.

west. With both the weather and the ice again closing in, the only prudent course was to set sail for Australia for the winter.

The next season Ross returned to the Antarctic hoping to extend his exploration east of the previous year's discoveries. Entering the pack at about 146° W, *Erebus* and *Terror* were stuck for fifty-three nerve-wracking days before breaking through to open water. They sailed on to the Barrier, but were able to extend their easterly penetration only seventy miles, their southerly record an additional six.

The glory belonged to the first of the two voyages: the discoveries of an active polar volcano, a major mountain range, and the massive face of the largest ice shelf on Earth (later named the Ross Ice Shelf) (Fig. 1.10). The coastal aspects of northern Victoria Land as far south as Mount Melbourne were sketched onto the map, but from there down to Mount Erebus, the mountains were an indistinct emanation viewed at a great distance. Most opportunely for those who would follow into the Ross Sea, the expedition had discovered the gateway to the interior, from which a half-century later heroic assaults would be staged on the Geographic South Pole.

The latter half of the nineteenth century was a period dominated by capitalist motives. Exploration for the sake of discovery gave way to ventures promising a profit. Sealers and whalers continued to ply the Antarctic margin, but governments and wealthy patrons were unmoved by the barren landmass, even in the name of glory through reaching the Geographic or Magnetic South Poles. One such would-be whaler was Henrik Bull, a Norwegian who had immigrated to Australia and was aware of Ross's reports of the sighting of many whales during his voyage of discovery. When he was unable to find backing to outfit a whaling vessel to the Ross Sea, Bull traveled to Sweden and made a pitch to eighty-four-year-old Svend Foyn, inventor of the explosive harpoon and designer of the first steam-powered whaler, and a rich man because of both. Foyn was enthusiastic about the proposition and offered one of his retired steam whalers, the *Kap Nor,* which was promptly fitted out and renamed the *Antarctic.* Foyn placed the ship under the command of Captain Leonard Kristensen, and designated Bull as his "agent." Throughout the expedition, the two wrangled, generating tension among the entire crew.

When the *Antarctic* reached Melbourne, a number of the crew jumped ship or were dismissed. One of the men to sign on as an able-bodied seaman was Carsten Borchgrevink, a schoolteacher of Norwegian birth, who was flush with the fever of Antarctica. Once on board he assumed the role of resident scientist and through his impetuousness was soon dining at the captain's table. The *Antarctic* left Melbourne on September 26, 1894, but because of a broken propeller it had to detour to New Zealand for repairs. After finally sailing south, the ship was stuck in the pack for thirty-eight days before breaking out into clear water on January 13, 1895. From there the captain set a course straight toward Cape Adare. The trouble was they were sighting no whales.

On January 18 they landed on a rocky beach at Possession Island. Borchgrevink went off exploring and came back with a sampling of lichens, the first plants to be found south of the Antarctic Circle, and a prize that had escaped the watchful eye of Robert McCormick a half-century before. From there the men sailed south, past Cape Hallett and Coulman Island (see Fig. 1.5). Bull was awestruck and perhaps intimidated by the scene of the

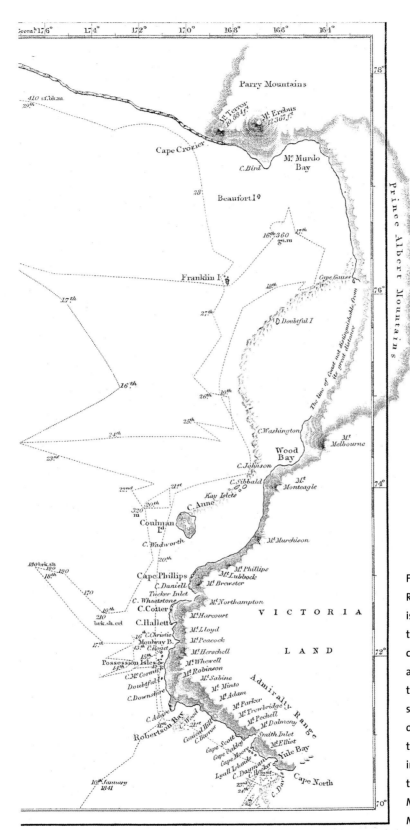

Figure 1.10. This portion of Ross's map of Victoria Land is drafted with south to the top. Note the detail of the coastal areas as far south as Cape Washington. From there to the south, Ross's ships stood far to the east of the Prince Albert Mountains, and detailed surveying was not possible. Note the presence of the Parry Mountains to the south of Mounts Erebus and Terror.

Admiralty Range: "The pinnacled mountains towering range beyond range in majestic grandeur under a coverlet of matchless white; . . . the sky of clearest blue and deepest gold when the sun is at its lowest; but perhaps more than all, the utter desolation, the awesome, unearthly silence pervading the whole landscape."

Still no whales. Returning to Cape Adare on January 24, they determined that the northern tip of the peninsula became a long cobble beach several hundred meters long. One of the whaleboats was lowered, and, after an hour of hard pulling, the sailors reached the shore. As they hit the beach, there was a great burst to be the first off the boat. In the aftermath, three different men claimed to have been the first to set foot on the continent of Antarctica. Captain Kristensen was in the bow and jumped as the keel struck land. He claimed that Borchgrevink jumped out second from the stern, but Borchgrevink claimed to have gotten ahead of the captain at the shore, to steady the boat for him. Likewise, a seventeen-year-old New Zealander, Alexander von Tunzelman, who had signed on at Stewart Island, claimed to have been first out of the bow to hold the boat.

Except for discovery of the first plants on the continent and accomplishment of the first landing on the mainland (tainted with acrimony), the expedition had been a disaster: no whales, few seals, and Bull and the captain always at each other. But as he sailed back on the *Antarctic,* Borchgrevink was hatching a bold plan, one that would return him to the ice, back to the beach at the tip of Cape Adare, and there to establish a base and to winter over.

The *Antarctic* landed in Australia on March 11, 1895. Four and a half months later, Borchgrevink was in London presenting a paper to the Sixth International Geographical Congress on his Antarctic accomplishments and his plans. The timing was critical. For several decades, momentum had slowly been building in England and in other countries for further exploration of Antarctica. One of the principal proponents was Sir Clements Markham, president of the Royal Geographical Society, and also chairman of the Geological Congress. At the meeting he promoted a resolution which read, "The exploration of the Antarctic regions is the greatest piece of geographical exploration still to be undertaken." Delivered with such raw enthusiasm, Borchgrevink's reported landing and his plans invigorated the discussion at the meeting. While Markham promoted a National Antarctic Expedition, Borchgrevink promoted himself. He proposed that he lead a party that would be dropped at the beach at Cape Adare with shelter and supplies to winter over. He reported that he had observed a route to the top of the Adare Peninsula that connected to the mountainous plateau, over which he would mount dog teams in an attempt to reach the Magnetic South Pole.

When official sources were unwilling to support this brash outsider, Borchgrevink turned to private donors. The following year he met Sir George Newnes, the publisher of a popular magazine called *Tit-Bits,* who provided Borchgrevink with £40,000 in exchange for the promise of exclusive rights to his story. Privately this commitment riled the members of the Royal Geographical Society, who by that time had managed to raise only about half this amount for the National Expedition.

By August 1898 an old, refitted whaling ship, renamed the *Southern Cross,* was sailing from London carrying thirty-one men and ninety Siberian huskies. The group included four scientists: Nicolai Hansen, a biologist; Hugh Evans, his assistant; Louis Bernacchi, a physicist; and William Colbeck, a magnetic observer, who together would be responsible for observing the weather, rocks, living organisms, magnetic field, and aurora.

The expedition was stuck in the pack for forty-three days before escaping to the north quite close to the point where they had entered. A second attempt farther east at about 174° E took only six hours crossing a light pack into open water of the Ross Sea. On February 17, 1899, the *Southern Cross* glided into an ice-free Robertson Bay, landing the party at the same spot that Borchgrevink had stood four years before (see Fig. 1.5). The Norwegian-Australian named it "Ridley Beach" to honor his English ancestry on his mother's side. Supplies were hurriedly unloaded, and two wooden huts were constructed. On March 3 the *Southern Cross* sailed north, leaving ten men and seventy-five dogs to face the polar night. The last penguins left the beach eleven days later.

Once the sea ice had frozen in Robertson Bay, various members of the party took short forays to its perimeter, but nowhere could they find a place to climb out, because steep cliffs and ice walls came down to the sea all around. Consequently, exploration was limited to the volcanic inner faces of the Adare Peninsula, to the impassable margin of the mainland, and to several small islands within the bay. On the western side of the bay, they did discover that the bedrock is composed of metamorphosed sedimentary rocks, whose layers were deformed into tight folds.

As total darkness descended and with no work routine, the party became slothful and began to grate on one another's nerves. Hansen fell ill in July, possibly because of scurvy (though none of the others showed signs of it) or to the residual effects of a malady contracted while passing through the tropics. His condition steadily deteriorated, until he died on October 14, less than a half-hour after the first penguin returned to Ridley Beach.

The ice had cleared from Robertson Bay by the first of the year, leaving the men anxious for the return of the *Southern Cross*. On January 27 the ship arrived. By February 2 it was fully loaded and the party was cruising south along the Victoria Land Coast. They managed a landing on the northwest side of Coulman Island, and then poked into the strait between the mainland and the island. Lady Newnes Bay, named for the wife of the patron of the expedition, opened to the south but was filled with thick ice tongues descending from the adjacent mountains (Fig. 1.11). From there the ship veered out from the coast, turned south, then sailed deep into Wood Bay to the north of Mount Melbourne (Fig. 1.12). Traveling back west about twenty miles along the peninsula that terminates at Cape Washington, they discovered a broad cobble beach and a cove that gave promise of safe haven for a ship were it to winter over. From there they sailed south into the outskirts of what later would be called Terra Nova Bay. An ice foot attached to the land south of Mount Melbourne appeared to extend all the way to McMurdo Bay, with the mountains beyond it of lesser elevation than the high peaks to the north.

About thirty miles south of Cape Washington, the *Southern Cross* tied up to the ice foot for magnetic measurements. Bernacchi climbed up to its flat top, where he glimpsed the polar plateau through a breach in the mountains. The emptiness of the scene released in him a deep shudder.

Figure 1.11. (opposite) Fitzgerald Ice Tongue pours from the Admiralty Range directly into Lady Newnes Bay. In contrast to the open water that Borchgrevink sailed in February 1900, this early-season photo finds Lady Newnes Bay frozen solid with ice from the previous winter.

Figure 1.12. Ross's expedition discovered Cape Washington (the dark termination of the peninsula in the lower left), Mount Melbourne (the volcanic cone to the right of center), and Wood Bay (covered by seasonal ice and extending out of the image to the right). Unusually open water in February 1900 permitted Borchgrevink to sail in along the coastline of Wood Bay for a distance of twenty miles. The upper reaches of Priestley Glacier appear in the left rear (compare with Fig. 7.11). Campbell Glacier silhouettes the left (south) side of Mount Melbourne, flowing right to left.

The sight that met our eyes was ineffably desolate. Nothing was visible but the great ice-cap stretching away for hundreds of miles to the south and west. Unless one has actually seen it, it is impossible to conceive the stupendous extent of this ice-cap, its consistency, utter barrenness, and stillness, which sends an indefinable sense of dread to the heart. There is nothing beautiful to contemplate, no contrasts, absolutely no diversity, but for all that it is majestic and affords a profitable theme for meditation.

The ship steamed on down to Franklin Island where a party landed and gave three cheers for Captain Ross. The coastline to the west remained as poorly charted as when Ross had visited the region sixty years before. After another landing on a small beach on the northern edge of Mount Erebus, the ship steamed eastward along the fringe of the Barrier encountering an inlet at 164° W. The captain moored the ship beside a low spot on the ice where a three-man party led by Borchgrevink was put ashore. With a dog sledge they headed south for a distance of about ten miles, thus recording two more firsts, reaching a new "farthest south" and becoming the first to mush dogs on the permanent ice of Antarctica, albeit a trivial distance.

Although the *Southern Cross* Expedition added little to what was known of the geography of the Transantarctic Mountains, it did accomplish a number of new landfalls and made geological, biological, and magnetic observations.

Meanwhile, the British National Expedition was slowly taking form. Following the International Congress, the Royal Geographical Society petitioned the government to dispatch a naval expedition, but the request was denied. In early 1898 Sir Clements Markham approached the Royal Society, headed by Sir John Murray, for cooperation in a joint venture. The Royal Society agreed, but with the caveat that the expedition be guided by clear scientific objectives, stressing in particular the importance of magnetic research. The public appeal was greatly aided by a gift of £25,000 from a wealthy donor in 1899. Then the Prince of Wales and the Duke of York became official patrons. Another approach was made to the prime minister, and this time the government promised £45,000 if that amount could be matched by the private sector. Sir Clements renewed his appeals with vigor and raised the money.

In March 1900 the keel was laid for a new ship to be built expressly for the expedition. It was constructed along the lines of English vessels that had evolved from earlier whalers. Reinforced for ice penetration yet worthy of the high seas, it had three masts and two strong auxiliary steam engines. The British favored this design over that of the *Fram,* the recently built Norwegian ship, which had a saucer-shaped bottom that allowed it to ride up out of the water if caught in tight pack ice, avoiding the tremendous pressures that potentially could crush even a reinforced hull. The drawback of the *Fram,* however, was that it rolled drastically in rough water, and the need to cross the stormiest seas on the planet to reach Antarctica boded for a miserable passage.

The Admiralty named the ship *Discovery,* the sixth British ship to be so-called, with her predecessors all involved in various aspects of Arctic exploration during the nineteenth century. The expedition was readied, the personnel selected, supplies and equipment were assembled, and the ship left England bound for New Zealand on August 6, 1901.

Commander Robert Falcon Scott, an officer in the Royal Navy with a keen interest in science, was Clements Markham's choice for expedition leader. Scott's leadership abilities, his vigor on polar marches, his eloquent writing style, and ultimately his tragic death were to place him at the center of the heroic pantheon of Antarctic explorers. The officers on the voyage were Albert Armitage, Charles Royds, Michael Barne, Reginald Skelton, and Ernest Shackleton. The scientific party included Louis Bernacchi, the physicist who had accompanied the *Southern Cross* Expedition; Edward Wilson, a physician, zoologist, and artist; Reginald Koettlitz, a physician and botanist; Thomas Hodgsen, a marine biologist; and Hartley Ferrar, a geologist.

Discovery put in at Lyttleton Harbour, New Zealand, where she was reprovisioned, her rigging refitted, and defects in the hull repaired, though an annoying leak persisted into polar waters. Scott recounted departure day:

> On Saturday, December 21, the "Discovery" lay alongside the wharf ready for the sea and very deeply laden. Below every hold and stowage-space was packed to the brim— even the cabins were invaded with odd cases for which no corner could be found. But the scene on deck was still more extraordinary. Here—again—were numerous packing cases for which no more convenient resting place could be found; the after part of the deck was occupied by a terrified flock of 45 sheep, a last and most welcome present from the farmers of New Zealand. Amidst the constantly stampeding body stood the helmsman at the wheel; further forward were sacks of food, and what space remained was occupied by our twenty-three howling dogs in a wild state of excitement.

Discovery sailed south from Lyttleton, under sail whenever possible to conserve the limited supply of coal. On January 3, 1902, the ship encountered the pack, clearing it with little trouble in five days. It sailed into Robertson Bay, with a landing at the hut on Ridley Beach. Bernacchi was the old hand, pointing out features, reminiscing, leading a party to Hansen's gravesite. From Cape Adare the expedition sailed south as close in as possible along the northern Victoria Land coast, but was forced out repeatedly by gales.

South of Coulman Island, pack and fast sea ice prevented them from entering Lady Newnes Bay, but to starboard the men could see the magnificent, icebound Mount Murchison with its twin summit, and the vertical, bare rock face of Cape Sibbald rising from the bay (see Figs. 1.5, 1.7). *Discovery* attempted to penetrate deep into Wood Bay on the north side of Mount Melbourne, to scout whether that reentrant offered a winter haven and a route to the interior, as Borchgrevink had suggested it might. Ice, however, foiled the plan, so *Discovery* sailed back along the peninsula, rounding the steep tip of Cape Washington (see Fig. 1.12). It was with keen interest that the expedition crossed into clear water south of the cape, for neither Ross nor Borchgrevink had been able to approach the land between there and McMurdo Sound.

The coastline south of Mount Melbourne took a broad swing to the west and then straightened out again and headed south. In places, a foot of fast ice stood out from the lower foothills; elsewhere, bare rock stretches of those foothills plunged straight into the sea. The mountains to the south of Mount Melbourne were less grand, with fewer peaks than those to the north and some with flat, tabular tops. At the lowest point on the skyline a broad glacier drained directly through a portal, giving a glimpse of the vast field of white that rose behind the mountains. In the middle of this mighty glacier was an odd, beehive-shaped nunatak—an island of land surrounded by ice—which split the glacier's flow and stood as a striking landmark in this rather subdued portion of the mountains.

Following the steep ice foot south, and watching the mountains unfold to the west, the men suddenly realized that the wall pinched out in a deep cleft, with the opposite wall striking back directly east. *Discovery* backed out and steamed east along its massive, jagged face for a number of hours with the height varying from 70 to 150 feet. After probably thirty miles the wall made a square right turn and headed due south. To all appearances, the ship had been following another of the barrier tongues like the ones in Lady Newnes Bay, but the scale of this one was gargantuan. Scott named this the Drygalski Ice Tongue, after Professor Erich von Drygalski, the leader of the concurrent German Antarctic Expedition (1901–1903) (Fig. 1.13).

The day was crystal clear as the ship rounded the corner of the ice tongue, with Mount Erebus 120 miles distant, puffing wisps of steam from its cratered summit. With Wood Bay out of the question that season, it was important for Scott to find a safe harbor for wintering over with the ship. The *Discovery* worked its way south through heavy pack, standing off maybe 30 miles from the coast, its men eyeing any hint of a headland that might signal shelter. On the second day, they saw a brown bluff or cliff that looked promising out in the distance, and set a southwesterly course for it. After several hours of rough pounding, even stopping the ship dead in the water on several occasions, they reached the mouth of an inlet and steered into it. Here was a perfectly formed harbor,

Figure 1.13. With a width of fifteen miles, Drygalski Ice Tongue debouches directly from the East Antarctic Ice Sheet into the Ross Sea through a narrow cleft in the mountains. Cliffs at its forward edge that are 150 feet high meet a thin layer of nascent ice as seasonal ice separates to the south (left). On January 18, 1902, Scott sailed in along the fast ice at the upper right of the image, at the perimeter of what Shackleton would later name Terra Nova Bay. *Discovery* found itself in a corner that forced a sharp turn to port (east). As the ship sailed along the north side of the ice tongue, the mast on *Discovery* was too short for the men to see over the cliffs. Edgeworth David's party crossed the ice tongue several miles from its release at the mountain front late in 1907 on its traverse to the Magnetic South Pole. The bight where the ice tongue leaves the shore, aptly named Relief Inlet, is also the approximate location where David's party rendezvoused with the *Nimrod.*

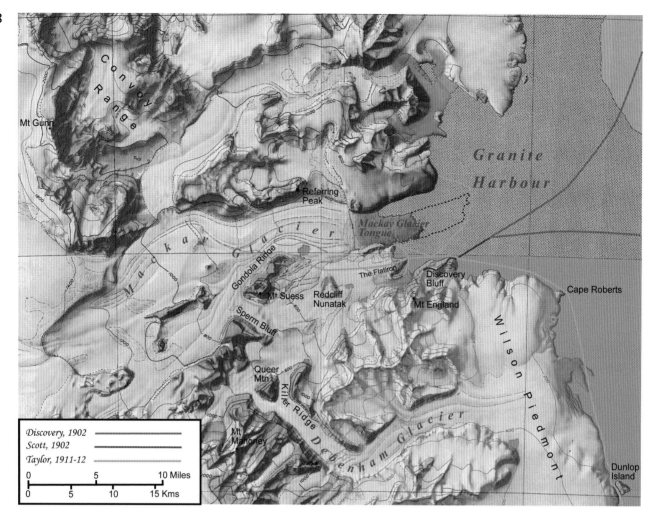

Figure 1.14. Shaded-relief map of Granite Harbour. Pink denotes the track of *Discovery*. When the ship had penetrated as deep as it could, Scott led a party of four on foot across the tight pack (red). When the men reached land at Discovery Bluff, they climbed up to its shoulder to view the ice tongue issuing into the center of the bay and the rocky walls rising around its perimeter. In 1912, during Scott's ill-fated southern journey, a party led by Taylor (yellow) retraced Scott's path onto Discovery Bluff and explored into the middle reaches of Mackay Glacier (see Figs. 3.13–3.16).

encircled by steep bluffs of dark red and brown rock, several hundred feet high, which converged on a glacier emptying into the head of the bay (Fig. 1.14).

The ice floes were extremely smooth and gave the appearance of having formed under quite protected conditions. They had only recently broken up, and were most peculiar looking in that they were cracked into perfect rectangles, unlike the typical polygonal floes ground or hoved up at their boundaries. *Discovery* pushed into the floes as far as possible toward a blunt cape on the south side of the fjord, and then stopped.

Scott organized a shore party consisting of Wilson, Koettlitz, and Shackleton, and soon they were scampering over the ice floes in their sea boots, crossing at points of contact and hopping across spaces of open water. At the narrow ice margin, fast to the foot of the bluff, a sea swell lifting the pack gave the men a final tricky step before they were on solid ground. Littering the base of the cliff were angular blocks of granite that had fallen from the outcrops above. This was the first granite found in outcrop in the Transantarctic Mountains, and Scott named the embayment Granite Harbour. The men scrambled along the foot of the steep cliff, found a route to climb to the top, and then hiked to the west along the crest. As they walked, they found orange lichens and clumps of green moss. The latter contained a primitive, wingless insect, *Collembola,* akin to the springtails that had first been discovered at Cape Adare by Borchgrevink's party. As they reached the corner of the bluff, the men were able to see into the southwestern recesses of the harbor and to look across to the ice tongue that separated them from the bluffs on the opposite shore (Fig. 1.15). The day was cloudless, water trickled down in rivulets from melting patches of snow, and the men lingered in the warmth of the sun there at their first landfall on the mainland. After collecting samples of rocks and the biota, they returned to *Discovery,* confident that they had found a safe winter harbor, if no alternatives could be found farther south.

From Granite Harbour, *Discovery* steamed south toward McMurdo Bay (Fig. 1.16). Although at first it appeared to be an open passage between Mount Erebus and the mountains to the west, closer in low lands became visible that seemed to connect across the expanse. Since no further penetration to the south seemed possible, Scott pointed his ship eastward out of the pack. Stopping at Cape Crozier, while others erected a message post in the prearranged center of the rookery, Scott, Wilson, and Royds climbed the loose scree of a 1,300-foot volcanic cinder cone behind the rookery. From the summit they were able to gaze across the Barrier, which presented an image of utter flatness, marked only by some low-amplitude undulations backlit by the sun. Both from there and along the Barrier, where low portions allowed a vantage from the crow's nest, Scott concluded that Ross had erred in his sighting of the Parry Mountains. Not meaning to impugn Ross's careful charting, Scott noted how deceptive appearances can be in this land, to the extent that "one cannot always afford to trust the evidence of one's own eyes."

About noon on January 29 the expedition sailed past Ross's easternmost point, keen to sight what he cautiously had noted as the "appearance of land." The air was clear, but they could see no land. The next day, however, as the ship sailed east, the ice gradually rose, and rock was sighted through a break in the fog that had settled in the night before. This new ground Scott named King Edward VII Land.

Returning along the Barrier, *Discovery* moored in an inlet at 165° 45′ W, 78° 30′ S, and several parties disembarked to sledge on the ice. The expedition carried a balloon named Eva to be used for aerial observation, which the crew hauled out on the ice shelf, tethered with a five hundred–foot rope, and inflated with hydrogen gas. Scott had "the honour of being the first aëronaut to make an ascent in the Antarctic regions." He observed the same broad undulations running parallel to the ice front that he had seen at Cape Crozier, but otherwise the view to the south was featureless. Shackleton, who got the second ride, went aloft with a camera, but a wind sprang up and prevented any other rides for the disappointed chaps in the queue. The spot where this activity took place was named Balloon Inlet, but it was later incorporated into the larger Bay of Whales.

Figure 1.15. This view of Granite Harbour is from the spot on Discovery Bluff first climbed by Scott, Wilson, Koettlitz, and Shackleton. This is also the place that Taylor's party left a depot that was crucial to the survival of Campbell's party on its return from the winter-over on Inexpressible Island. Emerging from behind The Flatiron, the buttress in the left center of the photo, Mackay Ice Tongue extends a short distance into Granite Harbour. From maps recorded during the heroic era, the ice tongue in the early twentieth century would have extended beyond the right edge of the photo. The dotted outline of the Mackay Glacier Tongue in Fig. 1.14 follows Taylor's map.

From this inlet *Discovery* retraced her track along the ice shelf skirting south along the west side of Mount Erebus. A long peninsula extended south-southwest from the volcano. Just around its tip the explorers found a small bay backed by fast ice and shoreline—the site of the present-day McMurdo Station. There *Discovery* moored on February 8. That night Scott weighed the pros and cons of this location for the winter quarters:

> From the point of view of traveling no part could be more seemingly excellent; to the S.S.E. as far as the eye can reach, all is smooth and even, and indeed everything points to a continuation of the Great Barrier in this direction. We should be within easy distance for the exploration of the mainland, and apparently should have little

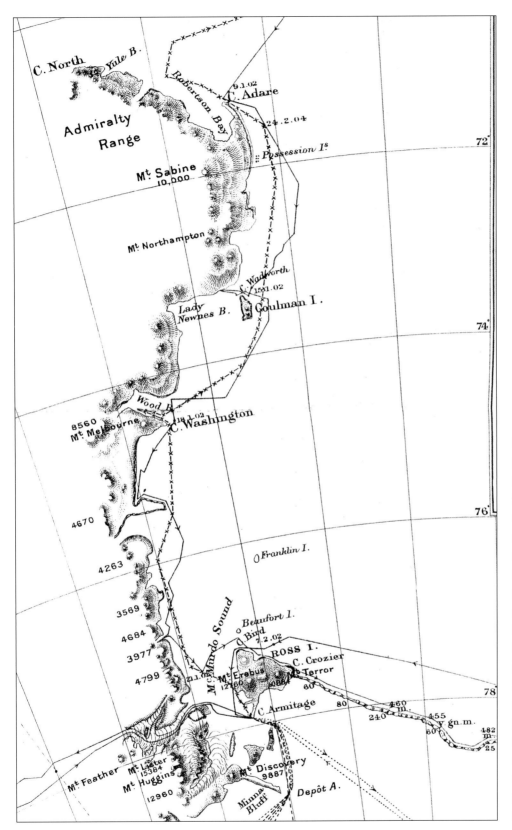

Figure 1.16. *Discovery*'s route along the Victoria Land Coast in 1902 is illustrated as a solid line in this map published by Armitage, the second in command, to accompany his account of the expedition. Note the landing at Cape Adare, the track into Wood Bay and out around the Drygalski Ice Tongue (unlabeled), and the track into Granite Harbour (at 3977)—all accomplished before Armitage sailed deep into McMurdo Sound for winter quarters near Cape Armitage. The x-marked line shows the return route of *Discovery* in 1904.

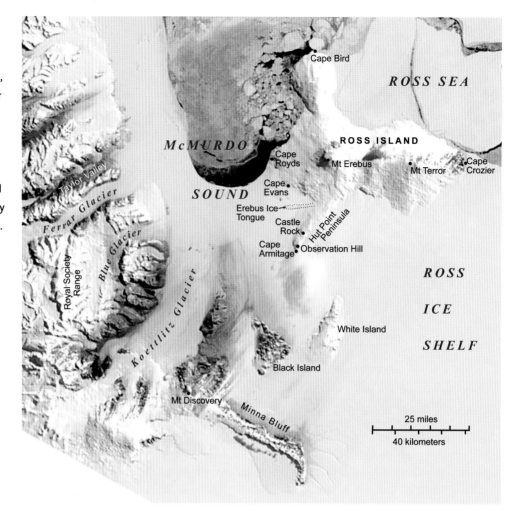

Figure 1.17. In this satellite image of the McMurdo Sound area, sea ice is fast as far north as Cape Royds, where Shackleton's winter quarters were located in 1907–1909. In 1902 Scott was able to penetrate as far south as Hut Point, whereas during his second expedition he reached only as far south as Cape Evans.

difficulty in effecting a land communication with our post office at Cape Crozier. There are no signs of pressure on the ice; on the other hand, the shelter from the wind is but meager, and one can anticipate intense cold and howling gales. On the whole to-night I feel like staying where we are.

The next day everyone set out to explore in different directions. When they reassembled that evening, there was great excitement detailing where each had been and arguing about the lay of the land as seen from different vantages. The two observations upon which they generally agreed were that the Parry Mountains did not exist and that Mount Erebus and Mount Terror formed an island, with the land that appeared to connect them to the mainland in fact being two low islands named simply White Island and Black Island (Fig.1.17).

Next the party set about to construct two small huts for geomagnetic observation and a larger one for storage. This square wooden structure with its low, pyramidal roof would also figure importantly in the later expeditions of Shackleton and of Scott, and today it is still standing on the outskirts of McMurdo Station, infusing the spot with the spirits of the forebearers. During the first weeks, the expedition named landmarks so familiar to those Americans and New Zealanders who work at McMurdo Station and Scott Base: Hut Point at the end of the spur where the hut was built and Cape Armitage at the tip of the peninsula for Lieutenant Albert B. Armitage, the second in command of the expedition. Behind this, Winter Quarters Bay, where *Discovery* was moored and supply ships for McMurdo Station dock today. Observation Hill was the name given to the prominent 750-foot volcanic cone beyond McMurdo Station that commands a superb view of all of McMurdo Sound. Behind this, Crater Hill, and between the two, The Gap, through which today runs the road that connects McMurdo Station with Scott Base. Arrival Heights designates the high cliffs along the northwestern edge of Hut Point Peninsula, and commanding the highest vantage on the ridge, five miles toward Mount Erebus from the hut, the bare orange faces of Castle Rock.

A number of short sledging parties, both man-hauling and with dog teams, set out to explore the area and to shake down the gear, in anticipation of more formidable campaigns to the south and to the west the following year. These first forays showed most of the men to be woefully inexperienced in dealing with polar conditions. Many suffered frostbite. Although individuals survived a number of near misses, one fatality did occur when a party returning from Cape Crozier was caught in a blizzard several miles up from the ship on Hut Point Peninsula. Rather than setting up camp, the men pushed forward with zero visibility. George Vince, one of the seamen, slipped on a precipice and plunged over Arrival Heights into the sea, where he was lost.

The sledge trip with the most important accomplishment was to White Island where Shackleton, Ferrar, and Wilson climbed 2,500 feet to the summit and were able to see far to the south. The coastline of the mountains met a high, volcanic peak (Mount Discovery), with a long ridge extending from it toward the east, where it terminated in a steep bluff (Minna Bluff). Beyond this the mountains could be seen to recede to the west and then continue southward into the murkiness of the horizon. But in front of this was smooth ice that connected through McMurdo Sound on back along the mountains as far as the eye could see. Here was the sure route to the south, with Minna Bluff chosen as the first major depot for the southern party the following summer.

2 From the Sea to the Ice Plateau
The Crossing of Victoria Land

Put yourself on Observation Hill, on a beautiful spring day in August 1902 (Fig. 2.1). The view is to the southern edge of the known world. The mountains to the west and the ice shelf to the south are the hinterlands of knowledge, with the boundaries left to imagination. The afternoon is a bit chilly at minus 45° F, but windless, so without the nip. The sun, still quite low in the northern sky, infuses the landscape with a glow rich in hues of pink and gold. This view is one of the grandest in Antarctica and a pilgrimage that every visitor to McMurdo Station must make (Fig. 2.2).

Across the sound forty miles to the west, the mountain range is a magnificent spectacle. Two specific features catch the eye: the graceful volcano to the southwest and the ghostly range of mountains directly across McMurdo Sound. The volcano, a perfect cone sweeping to nearly nine thousand feet, is a distant cousin of Mounts Erebus and Terror (Fig. 2.3). Scott will name it Mount Discovery, after his own stout ship. Extending

Figure 2.1. (opposite top) The 750-foot volcanic cone of Observation Hill rises behind McMurdo Station in January 1983. Open water in Winter Quarters Bay (right foreground) signals that the icebreaker has opened a channel for the supply ship. Pallets of retrograde cargo sit on the ice dock awaiting its arrival. Aside from the obvious settlement, the explorers of the *Discovery* Expedition would see that the terrain has been significantly altered since 1902 by the addition of a landfill on the forward edge of the base, scraped down from the surrounding hills. White Island marks the horizon to the right, with the Ross Ice Shelf to the left. Cape Armitage is the point at the right end of Observation Hill. The Gap is to the left of Observation Hill. Hut Point is out of the picture about two hundred yards to the right.

Figure 2.2. The Transantarctic Mountains extend along the western boundary of McMurdo Sound. The mouth of Ferrar Glacier opens into New Harbour, slightly right of center, below the rock wall laced with thin icefalls. To the left the broad breach in the mountains is the mouth of Blue Glacier. The Royal Society Range is just out of the photo on the left. Loose pack drifts in the sound in this late summer image taken from Castle Rock, five miles up the peninsula from Observation Hill.

Figure 2.3. Bathed in the glow of a midnight sun, Mount Discovery stands sentry at the far southwest quadrant of McMurdo Sound.

from Mount Discovery is a long ridge, maybe a thousand feet high and tens of miles in length, which ends in a steep bluff. It is geologically analogous to Hut Point Peninsula, on whose distal tip you are standing, each being a lengthy fissure along which volcanic lavas erupted. In front of the bluff are two low islands, White and Black, one mostly snow covered, the other not (see Fig. 1.17).

If you look really hard beyond Minna Bluff, you can see low mountains behind and to the south of Mount Discovery. They appear to diminish and then terminate at a large glacier. Shackleton, Ferrar, and Wilson also saw this view on their trek to White Island in fall 1902. From the end of the mountains a perfectly straight line marks the horizon across the entire southern quadrant (Fig. 2.4). The Parry Mountains that Ross had mapped do not exist. This was clear from the first day at Winter Quarters Bay, when Scott and Skelton hiked around the foot of Observation Hill and came back over the Gap between Observation Hill and Crater Hill. It may be that the ice shelf continues around the low end of the mountains and that they are in fact a major archipelago extending from Cape Adare to south of McMurdo Sound. This is what Scott had hoped to discover, because it would make for easy pulling toward both the South Pole and the Magnetic Pole. Through the long winter night, everyone on *Discovery* had been pondering what lay beyond the horizon, and they were anxious to explore.

Look back now to the mountains across the sound, to the Royal Society Range (Fig. 2.5). Ferrar estimated them to be upward of twelve thousand feet in elevation, with a summit ridge that is nearly horizontal. Dropping from the summit line is a massive, snow-covered wall partitioned by steep spurs, descending into an intricate array of foothills on the opposite shore. A short distance to the right is the end of a valley with thin glaciers pouring down its bare-rock northern wall (see Fig. 2.2). The embayment in front of this is New Harbour, seen from *Discovery* in spring 1902, when she steamed into

Figure 2.4. Level as the sea upon which it floats, the Ross Ice Shelf extends to the horizon to the south of Ross Island. White Island appears at the right of the image.

Figure 2.5. The twelve thousand–foot wall of the Royal Society Range looms across McMurdo Sound, with Mount Lister the summit at the right end of the ridgeline. Hidden from sight, Blue Glacier flows from left to right (south to north) behind the ice-free foothills and in front of the main range. Mount Huggins, ascended from the back side by Warren and Brooke during the International Geophysical Year (see Chapter 7), is the high peak to the left with the faceted face in sun. This image connects immediately to the left (south) of Fig. 2.2.

McMurdo Sound. Perhaps in this lower country the explorers would find a breach, or a portal, to the other side. North from there the mountains diminish in elevation and disappear over the horizon.

Now turn around 180°. Behold Ross Island in all of its splendor (Fig. 2.6). The summit of Mount Erebus is belching steam that rises in great puffs, drifting lazily to the north. The gentle incline below the summit drops off in a steeper middle section, which then fans out broadly at the base of the mountain. The peninsula you are standing on is about twenty-five miles long and aims like a rifle toward the shoulder of Mount Erebus, with Castle Rock, five miles off, sticking up on the ridge like the sight on a barrel.

Off to the right of Erebus is its twin, Mount Terror. Conjoined at the shoulders,

From the Sea to the Ice Plateau

Figure 2.6. From the summit of Observation Hill, Castle Rock, the highest outcrop on Hut Point Peninsula, aims like the sight on a rifle toward the shoulder of steaming Mount Erebus. The white gap between the dark segments of ridgeline closer in from Castle Rock is the saddle where Vince slipped and plunged down Arrival Heights on the far side of the ridge.

these flaring cones are mantled completely in white. To the south of the island is a snow field that even at this distance gives the appearance of a soft blanket. The party last fall, when Vince was lost, had attempted to reach Cape Crozier on the far side of Mount Terror, to leave a message for the support ship *Morning* in the "mailbox." They were seriously bogged down by the deep snow of this reentrant, which appears never to feel the force of winds. Had they made better time, they might have outrun the blizzard that caught them on Arrival Heights and swept Vince away.

If you were to jump over to the ridgeline to the north, you would see a narrow, elongate ice tongue issuing into the bay where Hut Point Peninsula meets the mountain. Its proportions and jagged outline remind one of the snout of a sawfish. It is now locked tight in the winter freeze, but last fall it floated freely in the lapis waters that opened northward and westward from the tip of the peninsula, with only the languid drift of loose pack to pattern its placid surface (Fig. 2.7).

Look to the bottom of the hill you are standing on. There in the little bay behind the point is *Discovery,* home to the explorers for the past seven months, and beyond her in the little saddle up from the cape is the square hut that served as storage all winter. There is warmth and shelter, there is food. Now sit back, dissolve into the scene, feel the distance and the silence, feel the cold.

The winter passed comfortably enough for Scott's party. Everyone was busy overhauling gear that proved woefully inadequate in the early forays the previous fall. Sledges, harnesses for the dogs, clothing, and food bags were all reconstructed. The group held regular theatrical performances, printed a small newspaper, and always celebrated holidays with special food and toasts.

Scott carefully laid plans for major sledge journeys to the south and to the west. He would lead the southern party, which would use all nineteen surviving dogs and would

Figure 2.7. In this late-summer view to the north from Castle Rock, seasonal ice drifts languidly out from the ice front. Outlined by its alternating light and dark edge, Erebus Ice Tongue crosses the photo in the near distance. The most distant point at the base of Ross Island is Cape Royds, site of Shackleton's 1907–1909 base. The bright white band immediately in front of that is Barne Glacier, and the thin, light line in front of that is Cape Evans, site of Scott's 1910–1912 base. The slope of Mount Erebus rises to the right. In the foreground is Arrival Heights, about a mile to the right (to the east) of where Vince perished during Scott's first expedition.

require a major commitment of support parties to place depots for the return leg. Armitage would lead a party of six to the west, seeking a passage through the mountains in the vicinity of New Harbour. A third party, led by Royds, was to attempt a crossing through the mountains to the southwest following the glacier that flowed out from the north side of Mount Discovery.

The spring saw a flurry of sledging trips in preparation for the major traverses. These were more and less successful, but the men were learning, in some cases the hard way, how to secure camp in a storm, how to cross crevassed terrain, how to handle dogs, and with each trip they grew fitter. Fitter, that is, until several of the men in Armitage's reconnaissance party to New Harbour developed symptoms of scurvy. This set all projected departure dates off by a month, during which time diets of fresh meat were ordered for the sick bay, and then eventually for all hands. The supply of fresh mutton ran out just as the first seals appeared along tide cracks, where they were promptly butchered. In the increasing light of spring, Koettlitz also raised messes of mustard greens and cress, experimenting both with hydroponic chemicals on flannel and with Antarctic soil, the latter providing the better yield. All men were served a ration of greens as an added boost against the scurvy, and those who had suffered eventually recovered.

When Royds's reconnaissance to the southwest ran up against deeply corroded ice overlain by morainal rubble at all turns, plans for the southwest party were changed. The new objective was for the advanced support party on the southern journey, led by Barne, to sidetrack on their return from depot laying and to scout the mountains and glaciers seen dimly to the south and west of Mount Discovery. When Royds successfully led a six-man party to Cape Crozier, planting the message of *Discovery*'s location so essential for the resupply of the expedition, the way was cleared for the launch of the southern campaign.

On October 30, 1902, cheered on by all hands, the twelve men of the two southern support parties left *Discovery,* hauling a train of five sledges loaded with 650 pounds of food for Scott's return. Scott, Shackleton, and Wilson pulled their dog-teams onto the trail three days later, with a number of *Discovery*'s crew and officers jogging along for a mile or two.

Armitage's recon to New Harbour had been disconcerting for more than the scurvy of his men. At their advance camp on the ice piedmont to the south of New Harbour, four of the party skied across the piedmont toward the mouth of the fjord inland of New Harbour. As they rounded the corner and looked to the west, a spectacular valley opened before them, with nearly vertical walls of bare, dark rock rising to more than three thousand feet on either side (Fig. 2.8). Above the north wall, snow-clad summits climbed upward to six thousand feet, feeding narrow hanging glaciers that poured gracefully over the shoulder and down steep, narrow gullies. These were the icefalls visible from across the sound at Observation Hill.

The fjord and the glacier that rose above it cut a straight trough through the mountains. Way up in the headreaches, as far as the glacier was visible, a blocky massif rose to seven thousand or eight thousand feet, with the appearance of a pass to its right that went through the mountains. This looked to be the gateway to the interior that Armitage was hoping to find. But the terminus of the glacier looked more than a little daunting.

Out from the glacier stood a huge moraine with mounds of ice-cored till in piles up to sixty feet in relief. Climbing over ice-cored moraine is a tedious business. Where sandy or gravelly rock debris is in contact with the ice, your foot usually has good traction, but you can't count on it. Loose boulders are often perched on the sides or tops of the mounds, and typically are poised to fall. It is like climbing a scree slope, two steps up, one step back, except that the scree is over ice, so three steps back is no surprise. Although one can climb over ice-cored moraines fairly safely, it is a frustrating business. To try to drag or portage a sledge across such terrain would be next to impossible. To the left of the lateral portion of the moraine next to the rock wall, Armitage also noted a frozen outwash stream twenty to one hundred feet wide that showed signs of melting at the peak of summer.

Figure 2.8. Ferrar Glacier, as seen from the mouth of the valley. The Kukri Hills line the right (north) side of the glacier, Knobhead is the distant peak, and Cathedral Rocks are the prominent buttresses in the middle ground on the left (south) side of the glacier. The rough ice to the left was what persuaded Armitage to seek an alternate route onto the middle portion of the glacier. When *Discovery* sailed past New Harbour in the summer of 1901, Armitage sighted a thin line of white that he thought might lead to the glacier. The route (in green) that he took down Descent Glacier is, however, in the reentrant behind the foremost, dark buttress on the south side of the glacier. The thin lines on the face of Cathedral Rocks were inaccessible to his party. The closest buttress on Cathedral Rocks is pictured in Fig. 2.11. Note the narrow icefall in the center of the face. The route that Scott forged up the glacier (in red) was along its south side, but the smoother north side was what he used on his return.

Beyond the moraine, the glacier itself was a raggedy mess. Weird, contorted pinnacles and blocks of ice, with relief up to twenty feet, rose from the glacier's surface. Sand and gravel and boulders were perched, ready to fall, from the tops of pedestals and shoulders, and were strewn across the pocked marked floor of the ice. The entire glacier seemed to be in a state of advanced decay. Melting during summer had obviously removed a great deal of the ice, but this early in the season everything was frozen tight. Armitage did not like what he saw—a field of grotesque ice traps with no promise of a route through its twisted maze. After the party's return to *Discovery*, this image would haunt his preparation for the western assault.

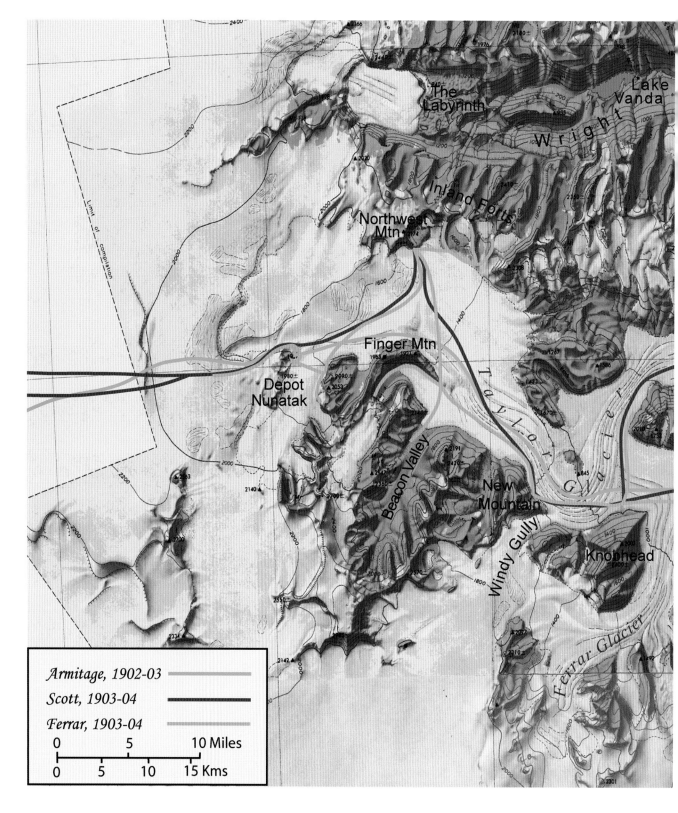

The Labyrinth

Lake Vanda

Wright

Inland Forts

Northwest Mtn

Finger Mtn

Depot Nunatak

Taylor Glacier

Beacon Valley

New Mountain

Windy Gully

Knobhead

Ferrar Glacier

Limit of compilation

Armitage, 1902-03

Scott, 1903-04

Ferrar, 1903-04

0	5	10 Miles
0	5	10	15 Kms

Figure 2.9. The topographic shaded-relief map of the Ferrar Glacier area shows the routes taken by Armitage, Scott, and Ferrar. Ferrar was the geologist in Scott's party who peeled off for geological exploration while Scott headed 250 miles onto the polar plateau.

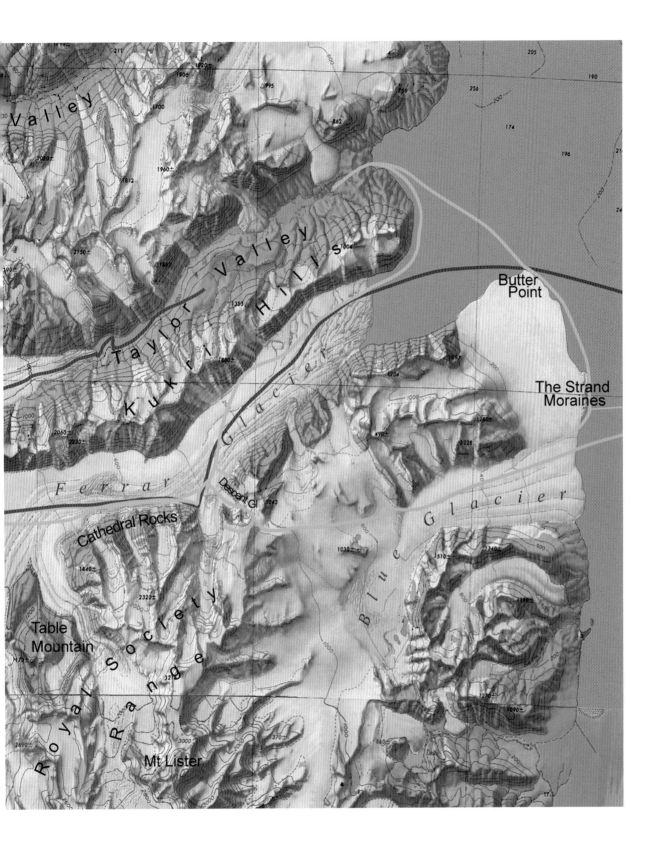

Valley

1920±
1806
1700

2080±

7812

1960±

2150±

7882

995

862

759

1004

1353

Taylor Valley

Kukri Hills

1880±

Glacier

2060±
2230±

Ferrar

Descent Gl 7243

Cathedral Rocks

1640±

2320±

Table
Mountain

2170±

3270

Royal Society Range

2690±

Mt Lister

370±

1030±

1000

370±

860±

1204

490±

Blue Glacier

1510±

1270±

1090±

2097

3228

1280±

1349±

1090±

256

174

196

205

190

200

211

24

Butter
Point

The Strand
Moraines

174

On November 29, 1902, the day before the southern party embarked, the western party was set to pull out from Winter Quarters Bay. With twenty-one men hauling ten sledges, this was the largest undertaking of the expedition. A, B, C, and D teams were divided between "main" and "supplementary" parties, each pulling a train of two or three sledges. Because of the alarming prospect of a direct assault up the glacier from New Harbour, Armitage's revised plan was to join the glacier by taking a roundabout route, running up to the northwestern headreaches of Blue Glacier (the one that drains the Royal Society Range), crossing a divide at the top, and from there descending to the main glacier along a steep slope of white that he had spotted from *Discovery* the previous year (Fig. 2.9). The men had named Blue Glacier for the color of the bare glacial ice in its lower reaches. The approach over this portion of Blue Glacier crossed alternating patches of blue ice and snow, and increased in steepness enough that Armitage's party had to relay sledges up one at a time, with the added annoyance of having to put on or take off crampons with every change in the surface.

In Antarctica, every gradation exists between soft white snow and hard blue ice. The metamorphosis from snow to ice occurs as fresh snow accumulates and old snow is buried. Snowflakes coalesce into ever larger ice crystals, giving the ice a granular texture, with the air space within migrating into a tightening network of bubbles. With enough time and pressure, the fine dispersion of bubbles in "white ice" is driven out or totally collapses, while the ice loses its granular texture and takes on a hue of blue. As glaciers move and are eroded or ablated at their surfaces, the older, deeper blue ice will find its way to the surface. Where crevasses open in blue ice they may be bridged by blowing snow, but unlike a bridged crevasse in an area of snow accumulation, which is very hard to spot, in a blue ice field the white bridges stand out in distinct contrast to the glacier ice, and so are easily seen.

After three days, the Armitage party reached a point where the main arm of Blue Glacier swings south and tributaries join from the west and the north. From there the going up the western branch was easier on a gently rising névé (snowfield) almost to the skyline, where Armitage expected they would find a pass to the line of white he had spotted the season before. On the evening of December 7, the party set up camp about a half-mile below the pass. Armitage, Koettlitz, and Ferrar skied up to the divide for the view into the middle reaches of the glacier (Fig. 2.10).

Armitage described the vista:

> A glorious scene suddenly opened into view. A typical glacier, having all the appearance of a river, lay some 2,000 feet below me. It appeared to be much crevassed, and to have rapids and falls of considerable extent. Opposite me, the grim-looking cliffs, which form the northern boundary of this glacier, extended east and west, gradually turning and terminating at northwest. Far to the westward could be seen a lofty range of mountains, 7,000 to 8,000 feet high, trending towards the northwest, and bordering the glacier on its southern side. The glint of the sun was on the ice, making the frozen river sparkle and shine like polished silver.

The view of the descending tributary to the main glacier was not glorious, however. In fact, Armitage decided that it looked too steep to attempt a descent.

That evening everyone crammed into the four tents of the main party for dinner,

Figure 2.10. On December 7, 1902, Armitage, Koettlitz, and Ferrar stood in the pass in the lower right of the image, looking down onto Ferrar Glacier and across to the corner of the Kukri Hills. The descent route looked so daunting that Armitage chose to try a route over the mountains to the west. This proved to be impossible (see Fig. 2.11), so the party returned to this spot and, following the green route traced on the image, lowered its sledges down the steep incline to Ferrar Glacier.

Figure 2.11. The eastern-most buttress of the Cathedral Rocks connects via the ridgeline to the summit of Mount Lister, at 12,209 feet the highest point in the Royal Society Range. The narrow platform at the skyline at the left edge of the image is the highest point that Armitage reached, where he looked down into the jumbled icefall with its open crevasses and upthrown seracs. From there he retreated to reconsider trying Descent Glacier, about a mile beyond the left edge of the image. The left (east) toe of the buttress was the location of one of Scott's depots during the 1903–1904 summer.

and then the support teams left for Winter Quarters Bay. Armitage now pinned his hopes of reaching the interior on doing an end run of the Royal Society Range on the snowy shoulder above the pass. This route wasted five days and a considerable amount of energy, because the slope became so steep that it required block and tackle to pull up the sledges. Beyond the crest of the shoulder Armitage had to confront a heavily crevassed and even steeper glacier than he had seen from the lower pass (Fig. 2.11). Beyond this a network of

GLACIERS AND CREVASSES

The Antarctic continent is blanketed by the largest sheet of ice on Earth. The transition to glacier ice from snow is one of compaction and recrystallization. The process begins with ice crystals (snowflakes) dropping from vapor-laden clouds, or, in the interior of the continent, directly out of clear atmosphere. At the atomic level, the repetitive union of molecules of water builds these delicate crystals. Although the forms that snow crystals may take would appear to be infinite, the one characteristic that they all share is hexagonal symmetry in their basic crystal-line structure. The snow crystals fall to the Earth, are blown across the surface and blunted, and eventually come to rest as tiny particles of ice that accumulate over time.

The average rate of accumulation (water-equivalent) for the whole of Antarctica is around eight inches per year. As snow is buried, the process of recrystallization begins. Because of slightly higher pressures at points of con-tact between the ice crystals, water molecules break free there, jump into the surrounding air space or dance along grain boundaries between the crystals, to points with less pressure where they recombine (recrystallize), building on the existing crystal structure. Larger crystals grow at the expense of smaller ones. The effect tends to segregate the ice from the air. The pervasive air space at the beginning of the process is reduced to interconnected tunnels and finally to discrete bubbles with no connectivity. By definition, glacier ice begins when the air space is no longer interconnected. At first the ice is white, but in time the air bubbles compress or seep out, and the purified ice assumes a deepening shade of blue.

"Like mighty rivers," glaciers flow down valleys under the force of gravity, but unlike rivers in which the water is a liquid (noncrystalline), glacier ice flows as a *ductile* or plastic solid (crystalline, albeit in motion). Flow rates for the major outlet glaciers of the Transantarctic Mountains have been measured in the range of two to six feet per day. If you sat on the side of a glacier and watched all day, the flow would not be quite perceptible. But if you set out some markers and surveyed them at the beginning and end of a field season, the distance that the glacier traveled would become clear. At the scale of the glacier, the movement is steady and uniform, but at the scale of individual crystals of ice, the frantic migration of molecules from spots of higher to lower pressure is causing the solid crystals of ice to recrystallize and change their shapes.

One of the characteristics of a ductile solid like ice is that, if the rate at which it flows is too rapid, then it will lose its coherency and break. The ice is then said to be behaving as a *brittle* solid. Because the ductility of ice increases with depth due to its overlying weight, ice is most brittle at the surface. As brittle fractures open they propagate downward, producing gashes in the glacier that are known as *crevasses*. Crevasses open where a glacier accelerates and stretches, and are oriented perpendicular to the direction of maximum extension (Fig. S.2). Where glaciers make sharp turns, flow over ridges, or drop from hanging valleys, the geometry of the crevasses may become chaotic, with huge seracs (ice blocks) tumbling down icefalls, stationary in human time.

Crevasses are like rattlesnakes—not a problem if you know where they are, but if you do not see them, they can catch you by surprise. The danger of a crevasse is that it may be covered by a bridge that conceals a yawning space below. A crevasse opens in tiny increments, with each brittle fracture separating the ice by a millimeter or so. As the crack opens, blowing snow sifts down into it, sealing up the gap and building a bridge that widens at the same pace as the opening of the crevasse. The bridge is typically thinnest at the edges and droops in the middle. To detect subtle crevasses, you need to look for faint linear offsets in the snow, and, if you find one, probe it with an ice axe or a pole to see how thin and wide it is (Fig. S.3). Then you must decide whether to cross or go around.

I first descended into a crevasse in 1970 about a mile out from our helicopter camp on McGregor Glacier. On an overcast day, I was belayed on a rope from above and climbed down a "crevasse ladder"—a flexible, wired ladder for crevasse rescue—into a world of deep, soft, and subtle gray. The crevasse was narrow and not more than six feet wide at the top. The walls reached twenty feet below to an irregular surface of blocks that had dropped years before from the underside of the bridge that we had chopped open for our fun. The paired walls undulated gracefully in symmetrical curves that transcended simple math, then

Figure S.2. Crevasses riddle the surface of Scott Glacier, one of the major outlet glaciers that cross the Transantarctic Mountains (see Chapter 6).

Figure S.3. Probing a bridged crevasse is the only way to tell whether it is safe to cross.

played off to the right into a slightly larger room that pinched to nothing at its bottom. I wedged my foot across the bottom of the crevasse and looked back up. I was surrounded by a sculpture illuminated from without. The walls were translucent gray, strewn through with layers of fine, white bubbles, configured in blocks that had been broken and then fused into a brittle/ductile, mish-mash of rehealed ice. I felt as if I were underwater—it was rapture.

I was hooked. Unless there was a particularly good 16mm movie showing at the camp that night, several of us would hike over to the crevasse field to fool around in our newfound world within the glacier. When the sun was shining, the grayness that I had experienced in my first crevasse was transcended by pervasive blue, pale and bright near thin spots in an overhanging bridge, dark and rich deep down in. The blue color is due to absorption of light in the red portion of the visible spectrum by molecules of water. It is the same in water and in ice. Crevassing is indeed an underwater experience. The deeper down one goes, the purer becomes the blue (Fig. S.4).

For the second half of the field season, we moved camp to the west side of Nilsen Plateau. Soon enough we found a promising crevasse about a mile from camp. It was a big one, marked by subtle sags and cracks in the snow surface, with its opposing sides separated by more than one hundred feet. How long the crevasse was we couldn't tell. I poked and chopped at the bridge along one of the sides and finally got an opening big enough to fit into. The apparent bridge of the crevasse, rather than spanning a hundred-foot opening, was a solid plug of ice as far down as I could see. It sat about three feet away from the glacier

Figure S.4. Crevasse interior, upper Scott Glacier.

wall, producing a narrow crevasse that was littered with blocks broken from the underside of its overhanging bridge.

Once I was down in the crevasse, I wormed my way laterally over a series of blocks, drawn toward a dark blue spot deep in the crevasse. After maybe fifty feet of this crawl, I came to the threshold of a gigantic room that opened abruptly. As I stood up on a big, wobbly block of ice, I gasped. At first glance I couldn't see any walls, only a deep, empty space of diffuse blue. What was this place?

As my eyes adjusted to the low light, I realized I was at the edge of a room at the termination of this broad crevasse (Fig. S.5). The walls were smooth and perfectly vertical and must have been more than one hundred feet high where they pinched together on the far side of the room. At that point the bridge was at its thinnest, glowing white high above. From there down to where I was standing, the bridge sagged in a graceful curve, its underside pocked by the loss

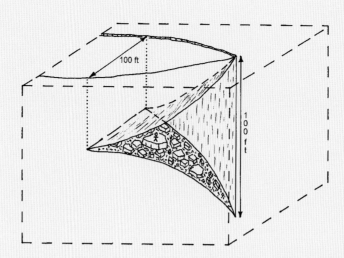

100 ft

100 ft

Figure S.5. Diagram of crevasse interior explored by the author in 1971.

of blocks that littered the floor of the crevasse. At that point the blocks met the bridge and merged into the plug of ice in the broad part of the crevasse that I had been crawling along to get here.

I signaled to be given more rope and climbed forward and down over a half-dozen of the big blocks till they dropped off steeply and I felt it would be unsafe to go farther. Standing close to the middle of the room, I could feel the immensity of the space, a nearly perfect tetrahedron, crafted by Nature, hidden away. It was like being in the hull of a giant ship, looking toward the bow. I lingered longer than I thought was fair, then crawled back to the surface so the others could descend into the blue. Although I have been in many crevasses since that time, each with its own fascination, there has never been another with such grandeur.

ridgelines rose to Mount Lister, the northern summit of the Royal Society Range. The party had to turn back if they were to find a route to the interior.

The next day a strong blizzard pinned them in camp, dropping several inches of wet snow, and the following day they were laid up in a thick fog. When the fog lifted, the party lowered its sledges down to the previous camp below the pass. Because of the extra resistance that the snow gave to the runners, and adamant that he would succeed in reaching the far side of the mountains, Armitage reconsidered lowering the sledges down the small glacier beyond the pass. Though one may question his route-finding abilities, Armitage's tenacity was dogged.

The morning of December 16 was fine and clear, although a fog hung part way down the valley. Armitage, Skelton, Evans, and Wild hiked to the pass to reconnoiter the route while the others packed up camp. By sighting down the slope, they measured that its average gradient was about 27°. Armitage put on a rope and attached a second to the nine-foot sledge that they had brought along. The others then lowered him down on the sledge while he probed for crevasses. Once the pitch proved to be safe, they anchored the sledge, the others hiked down in the tracks, and then they all repeated the operation several more times. After about six hundred feet of descent, the surface leveled out somewhat, giving an opportunity to regroup. The slope they had just descended may have reached 45° at its steepest, but from there down it appeared less steep, so the men hiked back up to camp to fetch the others and the loaded sledges.

They lashed the six sledges together in pairs. In addition to standard rope brakes, fashioned by tying a heavy rope across the underside of sledge runners, they also fitted bridles under the runners as brakes. Four men then carefully lowered the tandem sledges. Armitage, Skelton, Allan, and Macfarlane took the lead. Their run was touch and go when the sledges sped up and threw Armitage and Allan from their tow ropes, but the others managed to ride the sledges to a safe stop on the first terrace. Armitage made the cavalier comment, "It was a most exhilarating run, far more exciting even than the water-chute at Earl's Court." The other teams followed safely and from there maneuvered down two more inclines of lesser steepness, camping about a third of the way down the slope on a relatively level spot.

The next day saw the party to the valley bottom, where they congratulated them-selves on their success, and then were forced to lay up because of a thick fog that rose up the valley. When the air cleared the following day, the party quickly started to haul up the south side of the glacier, but the four inches of sticky snow that had been crucial to their successful descent now made pulling the sledges so grueling that only two could be pulled by the twelve men at a time. Furthermore, the snow made spotting crevasses dif-ficult, so Armitage was typically out in front probing the glacier with his ice axe.

By December 23 the party had pulled past the four faceted ramparts of Cathedral Rocks and reached a point where they could see that the glacier had two major branches, one coming from the south and the other from the west (see Fig. 2.9). Judging that the westerly route offered the better promise of reaching the plateau, Armitage pointed the party in that direction. By Christmas Eve they had risen over a shoulder in the glacier and then descended into a sort of null region, smooth and crevasse free. They crossed to the west of this névé and camped in the snowy lee of a pair of huge granite boulders set on a moraine that grew out of the foot of the blocky massif, the same mountain that Armitage had viewed in the spring from the mouth of the valley. Because of two prominent knobs at the summit, they named this landmark Knobhead.

From there on, the snow from the previous blizzard either had been blown away or had melted, so the pulling was easier. A few miles beyond their camp at Knobhead, the men noticed that a fourth arm of the glacier appeared to extend off and down to the north, but its full extent could not be seen. The mountains kept their secret, but only for another year. Over the next week between days lost to bad weather, the party ascended a series of rolling icefalls, separated by stretches of relatively smooth glacier ice. The up-permost reaches of the glacier gave the appearance of flowing down over large steps in the bedrock before channeling. From Knobhead upward the bedrock had changed from the steep walls of granite and gneiss (metamorphic rock) in the lower glacier to cliffs of flat-lying sedimentary rocks interspersed with thick sheets of dolerite. Although Armit-age noted that "their colouring was most beautiful, consisting of all shades of red, brown, and yellow," his overriding focus was the route to the plateau, and what might be beyond.

On the last day of 1902, with what appeared to be only one more icefall to the plateau, the party cached a week's worth of food at the last outcropping of rock, Depot Nunatak. But on New Year's Day, stormy weather again pinned them down. The next day, as they headed west, Macfarlane, one of the members of B team, collapsed. He was revived with some strong tea but clearly could not go on. Nor did other members of B team seem very fit, so Armitage ordered the party to camp while he and the A team pushed westward onto the plateau, vowing to return within five days.

Having mounted the final icefall, the party at last looked out across the endless plain of white that is the polar plateau. If there was a western shore to Victoria Land, it lay far beyond the horizon. Armitage and his party had forged the first route through the mountains, stood at the shore of this inland sea of ice, shivered at its vastness, and felt the puniness that every sailor feels when adrift beyond the sight of land.

As the party was approaching the campsite of the B team, stepping over narrow crevasses, Skelton mentioned to Armitage that he thought he could make out two figures. As Armitage looked up, he stepped into a crevasse and fell seventeen feet, hit an ice pinnacle on one wall with his thigh, and bounced to the end of his harness twenty-seven feet down. At that depth the crevasse was four feet wide and "widened out into what appeared to be a huge fathomless cavern." The others quickly lowered a rope and hauled out Armitage, who escaped with no broken bones and only his wind knocked out. Armitage had glanced up for only a moment, and the crevasse had surprised him. One advantage of man-hauling over other forms of traversing is that one is tied to a sledge by means of the harness rope and thus caught if a fall occurs.

After the A and B teams regrouped, their return to *Discovery* was uneventful. At Cathedral Rocks, they came upon a torrential meltwater stream between glacier and rock, where they camped to the sound of rushing water that could be dipped without having to melt snow. The climb up Descent Glacier was accomplished in eight and a half hours, relaying one sledge at a time, and using block and tackle on the steepest slope at the top. The stoic Macfarlane, who probably had had a heart attack (though it was not diagnosed as such at the time), climbed Descent Glacier one slow step after another, but had to be carried back on a sledge for most of the inward journey. After fifty-two days on the trail, the tired men were given a rousing welcome as they approached *Discovery*, then a bath, clean clothes, and food when they went aboard. Everyone was keen to hear the stories of their encounters with new lands where a mighty glacier cut clean through the mountains, descending from an ice plateau whose farther limits could only be imagined, and to offer warm congratulations, as Armitage said, "to the first human beings to storm and carry the heights of South Victoria Land."

After the successful return of the western party on January 18, everyone on *Discovery* was pondering the whereabouts of the southern party and, as the days went by, anxiously awaiting their return. On February 3 Skelton and Bernacchi were about 6 miles out on the sea ice when they saw a bedraggled threesome plodding toward them. Scott, Shackleton, and Wilson had traversed about 250 miles to the south, and with great hardship had barely managed to make it back alive. All three were suffering from scurvy, Shackleton in particular, who had broken down toward the end of the journey and had to be carried on a sledge.

The welcome back at the *Discovery* was a gala affair with all flags flying. As the southern party soundly shook the hands of the well-fed, tidy crew, they finally gained a perspective on their deplorable condition. Wilson noted, "I began to realize then how filthy we were—long sooty hair, black greasy clothes, faces and noses all peeling and sore, lips all raw, everything either sunburnt or bleached, even our sledges and the harness—things one didn't realize before, and our faces the color of brown boots, except where the lamp soot made them black." A bath and clean clothes preceded a huge dinner, and then the weary explorers retreated to their bunks with letters from loved ones, and finally the relief of sweet sleep.

The good news had been that *Morning* had arrived a week before with resupplies and a year's worth of dispatches from the civilized world. The bad news was that she lay eight miles out at the edge of fast sea ice, which was showing no signs of breaking out. As February passed, the situation remained unchanged, making it finally clear that *Discovery* would have to remain for another winter. Supplies were transferred, a cache of coal was left, and *Morning* sailed north on March 1 carrying eight volunteers from the original *Discovery* crew and Shackleton, whom Scott had ordered to return because of his bad health. Shackleton had protested bitterly that base food and a winter's rest would see him right, but Scott had been unmoved.

Winter passed quickly. Life was easier owing to the experiences of the previous year, and more pleasant because the few malcontents were all among the departing volunteers. When the sun returned on August 21, the men were eager to go out sledging. A number of short sorties filled in the map around McMurdo Sound. Royds and Wilson with four others sledged over to Cape Crozier in mid-September, hoping to collect some Emperor penguin eggs at the stage of partial incubation, but they were startled to find that the chicks had already hatched.

The major undertaking of the season, led by Scott, was to be a sledge journey into the interior of the ice cap, following Armitage's route through the mountains. Sandstone specimens that he had collected raised the possibility of finding fossils in these sedimentary rocks, thereby enabling the dating of the sequence. If found, fossils would be a scientific first for Antarctica. Beyond this, Scott hoped that although the ice appeared to continue to rise beyond Armitage's westernmost penetration, it would eventually fall off to a distant shoreline beyond a western slope. If there was another side to the plateau, Scott would be the one to find it. But even if there were not, taking the measure of such a vast ice sheet would itself be a major geographical accomplishment. Scott led a six-man reconnaissance party, including Skelton, Dailey, Evans, Lashly, and Handsley, across to New Harbour with the hope of finding access to the lower portion of Ferrar Glacier, thereby avoiding the precipitous route over Descent Glacier that Armitage had forged the season before.

On September 14 the party spent all day working its way in along the south side of New Harbour, grappling with disturbed sea ice. That night the men camped at the terminus of the glacier, several miles up the fjord, with the sun casting only a faint crimson glow in the sky above the darkened valley. The air temperature was a frigid minus 49° F, but in the morning the sun was bright and spirits were high. The fjord was a sight to behold (see Fig. 2.8). Unlike the floating glacier tongues of northern Victoria Land that projected out from the mountain front, this floating glacier had calved well up into its valley. The walls on both sides, which rose steeply to more than three thousand feet, were devoid of snow and plastered over with morainal debris, except for narrow icefalls that poured into some of the gullies from the snowfields above.

The glacier was a raggedy mess, just like Armitage had reported, but that frozen stream along the edge of the glacier looked like a possibility to Scott. While the others climbed up the side of the valley to see whether any routes could be spotted from above, Scott and Skelton went to explore up the draw. The bottom was very slick in places, rough with gravel in others, but from what they could see, not too steep or too narrow for managing a sledge. About a mile along, the stream channel headed up onto the glacier. As the two men climbed, the relief on the ice pinnacles and blocks became more subdued, giving promise of what lay ahead.

When the groups joined up again, the men who had been up the side of the valley confirmed that it seemed to smooth out farther up glacier, and maybe that there was even a better route closer to camp. The rest of the day was spent dragging and portaging the two sledges up the twisted gully, with the party camping in a relatively smooth spot for the night. The following day they relayed the sledges for another half-mile, then pulled on up together on a smooth and easy incline to the eastern toe of Cathedral Rocks (see Fig. 2.11).

It had taken Scott's party six fewer days to reach this point from *Discovery* than it had Armitage's party the previous year, even with a day lost to a blizzard on the sound. Although they had to carry the sledges across some rough stretches of ice, the dangerous traverse down Descent Glacier could be avoided. On the moraine at the foot of Knobhead, next to a conspicuous, white, quartzite boulder, Scott made a cache: three weeks' rations for six men, four gallons of oil, and sundry climbing gear. This would be an easy place to find when the western party pushed through several weeks hence.

Upon their return, the party avoided almost all of the rough terrain by taking a more northerly route down the glacier. Scott surmised that the difference in the surface roughness resulted from midday sun from the north casting a shadow on the north side of the valley while warming and melting the glacier on the south, in contrast to the wee hours when the sun was in the south and at a lower angle, when its rays did not readily melt the north side of the glacier. From their camp at Butter Point, Scott pushed the party to test their limits by sprinting the forty-five miles back to the ship in two days.

The second major undertaking of the summer of 1903–1904 was a traverse to the southwest, led by Barne, into the reentrant south of the bluff, to explore the deep cleft into the mountains, with the possibility of finding another route to the interior. A party went out in mid-September to lay a depot for this traverse on the south side of White Island. Their thermometer broke at minus 67.7° F, nearly 20 degrees colder than Scott's party had suffered during the same cold snap. On October 6 the twelve members of the southwest and southwest support parties pulled out of Winter Quarters Bay.

Scott's western party, also numbering twelve, left the base on October 12. With the wisdom of the previous season, sledge groups had been scaled back. For this journey, the men were divided into two teams of six, each pulling a pair of eleven-foot sledges in train. Making excellent time, the party passed the depot at Cathedral Rocks on October 16. The following day they were on the moraine down from Knobhead, having taken six days to reach the same spot that Armitage's party had struggled to in twenty-seven days the season before. Over the next two days, however, the runners on three of the four sledges split or lost their coverings of German silver. Faced with such a disastrous breakdown, Scott placed most of the supplies and the one good sled in a depot and beat a rapid retreat back to *Discovery*.

The three sledges that the party brought back were beyond repair, but salvaged parts from two of them were rebuilt into a seven-foot sledge, and another eleven-foot sledge not being used was pressed into service. On October 26 the western party set out again, much more lightly provisioned. This time they were nine men, three bound for the plateau (Scott, Lashly, and Evans), three in support (Skelton, Feather, and Handsley), and a geological party of three, including Ferrar, Kennar, and Weller. The six bound for the summit pulled the larger sledge, the geological party the smaller one.

Everyone was in top shape from all the pulling on the aborted journey, and the loads were considerably lighter, so by the end of the second day, the party was already camped next to Descent Glacier. The runners on the sledges began to fail again, requiring makeshift repairs as they proceeded. When the men reached the depot of their previous traverse on November 1, Scott realized with horror that the instrument box had been blown open by the wind, and the "Hints to Travellers" manual for calculating longitude and latitude from celestial bearings had been lost. This book was critical for the plateau party once it reached the featureless plain of the summit, beyond any landmarks that could be used for navigation. Scott, however, decided to risk pushing on, in hopes that on the return the men would be able to retrace footprints, and that taking readings of the sun's noon attitude would give them a latitude along which to make their return.

The headreaches of this outlet glacier open into a broad basin rising through a series of steplike icefalls. Katabatic winds regularly pour down from the plateau over these upper slopes and through narrow side valleys, clearing much of the snow from the icy surfaces of the glacier. (Katabatic winds result when air descending from the stratosphere over the interior of Antarctica is cooled by the ice cap, becomes more dense, and pours down the incline of the ice. These gravity-driven winds funnel into the heads of outlet glaciers throughout the Transantarctic Mountains, creating a draft that is constant unless disturbed by encroaching storms.) Because of the lack of snow at the edge of the plateau, it became a problem for Scott's party to find surfaces on which they could secure their tents.

This condition was especially influential on November 4, when under increasingly forceful winds, the men pushed up over the undulation of one of the last icefalls hoping to find flat ground at the top. The surface did flatten, but no snow could be found over the ice. After about an hour of desperate searching, with everyone suffering frostbite on the face, a patch of white presented itself on the blue. It then took another frantic hour under gale-force winds to securely erect the three tents. At the time the men were happy to be out of the wind and in the relative comfort of their tents, but little did they realize that they would be pinned down by this howling blizzard for a full week, testing every man's ability to persevere.

Imagine what it was like in Scott's tent near the end of the week. Scott, Evans, and Lashly were in their one-man sleeping bags, their noses barely poking out. The flapping of the tent was so loud that one had to shout to be heard. But no one was talking. Conversation had been lively at the beginning of the week, but the usual subjects soon wore thin. Scott had a copy of Darwin's *Voyage of the "Beagle,"* and readings aloud had been amusing for awhile, but eventually everyone was bored of that, too. The high points of the day were when the men rolled up their sleeping bags and lit the primus stove for the morning and evening meals. It gave one something to do other than lie in the sack. These were also the times for which one waited to answer the call of Nature. When someone was out, there was a chance to hail the others inside their own cocoons to see how it was going. Otherwise, no communication occurred between the tents.

Each time the stove was lit, however, some melting occurred from the frost on the inside of the tent that collected from the men's breathing and from the snow that continuously seeped in through small holes or burst in when the door was open. Add to this the men's normal perspiration in the bags wetting them from the inside. Much as they tried to avoid it, moisture was soaking into the bottoms of the sleeping bags, and the men were beginning to find it hard to keep their feet warm. Even aside from cold feet, a man could force only so much sleep, and then he had to lie there conscious, wishing for some activity that would stimulate the atrophied muscles and bring concentration to a listless mind.

At that moment each man was just staring off at some unfocused point on the ceiling, and Scott was thinking to himself that regardless of the weather the next day, they should make a run for it.

They did attempt the next day to dig out the sledges and supplies buried deep in snowdrift, but wind quickly brought on frostbite, so the party waited out one more day. On November 11, with the wind only partially abated and visibility still "thick as a hedge," the western party broke camp and headed off in different directions. Scott and his men trekked west onto the plateau, while the geological party turned east to examine the rocks up close.

This was the moment that Ferrar had been waiting for. As a member of the Armitage D team the previous season, he had only glimpsed from the top of Descent Pass into the outlet glacier that would come to bear his name. All the way up this season he had been watching the mountains on both sides and examining the loose rocks in the moraines. The main rock types came down to four: a coarsely crystalline granite, both pink and gray varieties; metamorphic rocks, mainly banded gneiss (although Ferrar also had collected marble around Blue Glacier the previous year); sandstone, a rich yellow color, mainly rounded quartz grains; and dolerite, a dark, fine-grained igneous rock.

The explorer discovering new lands demonstrates those discoveries with a map of terrain showing the mountains and drainages. The geologist fills in this map by showing the distribution of rock types throughout the terrain. For a geologist like Ferrar, he was both making the map and filling it in with geology.

In the lower reaches of Ferrar Glacier the deep walls were composed of granite and gneiss, but midway up the glacier, in the vicinity of Cathedral Peaks, a dark horizontal layer began to appear overlying the granite high up the slopes (see Fig. 2.11). Ferrar presumed that this layer was the dolerite of the moraines, but he had not yet seen it up close in outcrop. When the party reached Knobhead, it had found dolerite right down to the base of the mountain, and no granite in sight. Partway up Knobhead were horizontal layers of sandstone, then at the top another thick sheet of dolerite.

Two possible scenarios presented themselves to Ferrar as he ascended the glacier. The horizontal contact between the granite and the dolerite suggested that the top of the granite was an old erosion surface (an unconformity). But had the dolerite been erupted as a layer of volcanic magma over this surface, with subsequent deposition of the sandstone, and then more volcanism over that? Or, alternatively, had the sandstone been deposited over the granite, followed by *intrusions* of the dolerite along the erosion surface and between sandstone layers, a type of tabular pluton called a sill?

That question had been answered as the combined parties passed Finger Mountain a week earlier (Fig. 2.12). There, magnificently displayed on its vertical face, was an outcrop where the dolerite cut across layers of sandstone stepping from a lower horizon to a higher one. Blocks of sandstone hung in the once liquid dolerite, dislodged as the magma intruded though it. This showed clearly that the dolerite was an intrusion formed later than the sandstone.

As soon as the western party broke camp, the geological party pulled about three miles down into the windless, sunny lee of Depot Nunatak (see Fig. 2.9). This was a place to dry out all the sleeping bags and clothes, so while Kennar and Weller spread them out across the moraine, Ferrar hiked up to the nearest outcrop. The rock was all dolerite with weathered surfaces a rich, chestnut brown, belying the dark gray coloration of the fresh rock. At this locality the dolerite formed exceptionally well-developed columnar jointing. Caused by the cooling of the sill, which contracted and fractured into a honeycomb of smooth, flat-sided columns, the outcrop was reminiscent of the Giant's Causeway in Ireland or Devil's Postpile in the Sierra Nevada. Ferrar found no sandstone in the outcrop there, but back down on the moraine were sandstone boulders with intriguing black flecks in them, possibly graphite, possibly of organic origin. The day had been a good start at exploring the geology.

The next day the wind had dropped off completely and the sun was shining in a cloudless sky. The men pulled across the blue-ice surface to the cliffs on the south side of Finger Mountain (see Fig. 2.12). The rock face loomed, perhaps three hundred feet of sandstone overlain by five hundred feet of dolerite. The cliff was fairly sheer, but a steep spur gave a path for climbing up through the section. Ferrar donned his tools of the trade, a rock hammer, compass, and sample bags, and started climbing.

As with any sedimentary sequence, the first thing that struck him was the layering, straight and horizontal, but within individual beds were the wispy festoons of crossbedding, telling him that water currents had worked these sands. The grains were essentially pure quartz, coarse-grained and even granular at places. Here and there were thin layers or lenses with quartz pebbles, in rare cases as big as hen's eggs. It would have taken some pretty strong streams to move cobbles that large, currents that had swept away any

Figure 2.12. A somber day at Finger Mountain. Here is where Ferrar determined that the dark dolerite he had been finding on the moraines was an intrusion rather than a lava flow. As can be seen, the dolerite magma intruded along a bedding plane of the light-colored Beacon sandstone, producing a sill, the thick central band of the outcrop. Fracturing during cooling of the sill produced columnar jointing, the vertical grain visible in the face of the sill. A large offshoot of the sill diked up along the right skyline and then flattened into a higher sill (the bottom of which can be seen as the flat bottom of the triangular face at the shoulder to the left of the summit). A graceful arcuate dike shot off the bottom of the sill, raising slivers of sandstone only slightly before the magma solidified.

matrix of mud. When one is looking for fossils, often the best rocks for preservation are the ones made of mud, the shales, which come to rest in low-energy environments. But these sandstones, from what Ferrar could see, had no shaley interbeds.

Then Ferrar let out a whoop. About one hundred feet up the spur, he found a thin, black seam in the sandstone, maybe one-quarter of an inch thick, running maybe twenty feet out the face. He broke out pieces of the sandstone along the seam and split them with his hammer, making quick work of all the rock that could be easily pried loose, but no luck. The black material looked coal-like, the sort of deposit that would come from organic material, plants probably, but no impressions of leaves or other possible fossils could be seen. He guessed that heat from the overlying dolerite sill had cooked these rocks to the extent that any vestiges of fossils that might have been there had been destroyed.

Nevertheless, this discovery was tantalizing, and gave hope that somewhere in the upper drainage of Ferrar Glacier a fossil would be found.

In 1902 fossils were still the only way to date a rock and place it on the geological timescale (Appendix 2). Bertram Boltwood would not perform the first radiometric date by isotopic analysis for several more years. So for Ferrar, a fossil was the grail that would allow him to relate his sequence of sandstones to those rocks elsewhere around the world that had formed at the same time in Earth history.

From the campsite on the north side of Finger Mountain the party pulled over to the north side of Ferrar Glacier and camped at the foot of the Inland Portals at the same site where the outward party had camped on November 2. For four days Ferrar crawled all over the bare-rock ridges of the Inland Forts, including an eight hundred–foot climb up one of the steep cirques, but he found no fossils and no more carbonaceous rock. On November 18 the party crossed back to the moraine downstream from Finger Mountain and from there hiked deep into Beacon Valley, the ice-free canyon for which Ferrar named the sedimentary sequence Beacon Sandstone (Fig. 2.13). Still no fossils. On November 21 the men moved down to a camp on the moraine at Knobhead, having picked up supplies at the New Mountain depot. For the next several days Ferrar worked around the south side of Knobhead, climbed eight hundred feet up its eastern face, and hiked across the glacier to Solitary Rocks. On the morning of November 24, he hiked over to the northern tip

Figure 2.13. Horizontal layers of Beacon Sandstone constitute the walls of Beacon Valley, first explored by Ferrar in January 1903. This was the first place where the horizontal sedimentary rocks of the Transantarctic Mountains were named. With the recognition by later geologists that these sedimentary rocks occur throughout the length of the Transantarctic Mountains, the name Beacon Supergroup was applied to the entire suite.

of the Kukri Hills. The northern arm of the glacier appeared to drop away toward the northeast, but as Ferrar rounded the corner to see what happened to it, a cloud rose up from below, obscuring his view. Once again, the valley held its secret.

That afternoon the party broke camp and moved on down the glacier into granite country, leaving the sandstone outcrops behind, and with them any hope of finding fossils this trip. Four more days and two camps saw them off the glacier, and pulling north across New Harbour to the mouth of the valley to the north of the Ferrar Glacier drainage. The night of November 29, the party camped in the mouth of this valley.

Ferrar was looking for what happened to the northern branch of the glacier, lost to clouds higher up. The mouth of the valley was bare rock, littered with boulders and stones mainly of granite and gneiss. About two miles up, an imposing piedmont glacier (now known as Commonwealth Glacier) flowed down the northern wall and spread almost entirely across the valley floor (see Fig. 2.9). A piedmont glacier is one whose ice spreads across a relatively flat area from some source higher up. It can be broad like the Wilson Piedmont Glacier (see Fig. 1.14), or small and defined like Commonwealth Glacier, which assumes a particularly circular form and vertical walls rising more than fifty feet. This is what blocked Ferrar's view up the bottom of the valley. It was frustrating. Only the bare-rock higher elevations showed beyond. The question was, did the northern branch of the glacier peter out somewhere up this valley, or did the valley head in a high divide with the northern glacier stopped behind it? It felt like the valley should continue through the mountains, like the adjacent outlet to the south, but the proof would be in the observation. Ferrar did not have time to climb farther up that afternoon, but in the morning he set out to skirt the southern end of Commonwealth Glacier. Just about the time he reached the breach between the glacier and the southern wall of the valley, a thick snow squall came down and obliterated the view once again, leaving the question hanging, what was there?

The geological party worked its way back to the mouth of Blue Glacier, with Ferrar suffering a case of snow-blindness that lost the group several days. On December 10 the party started the thirty-mile crossing of McMurdo Sound, and after a strong showing in the morning, pushed it on through, arriving at *Discovery* by 10:00 P.M. on the same day.

When Scott's western party escaped "Desolation Camp" on November 12, they headed out onto the plateau, where for two and a half weeks of pulling, interspersed with storms, they continued across the featureless plateau of white. Temperatures were markedly colder and the air thinner than the party members had experienced the previous year on the ice shelf. On the prearranged date of December 1, they turned back, having determined that the ice sheet extended at least 150 miles to the west of the mountains, and at that distance no western shore to Victoria Land was anywhere indicated.

Victorian images of geographical discovery typically portray the explorer in some fantastic landscape, mountains rising from the mists, canyons disappearing into a dark abyss, a jungle, deep and rich with detail. The Transantarctic Mountains are a fit setting. In complete contrast, the exploration of the East Antarctic Ice Sheet evokes a scene of starkness: a field of white, the horizon, and the sky. The explorer, not the landscape, becomes the subject of the frame. Without landmarks to offer distance, the horizon contracts. This is true on the ocean or the ice sheet, but from the crow's nest of a ship, one hundred feet above the water, the horizon will be 13.5 miles distant, whereas a man on foot will see only about 3 miles. In reality, the popular notion that on the plains one "can see forever" is not borne out. A sky full of moving clouds may give a sense of distance, but on the polar plateau on a cloudless day, or even worse on a gray day, the walls close in on an image of monotony and desolation. Scott, Lashly, and Evans had furrowed deep into this monotony and found their way back. The blank space on the map remained blank, save for a thin, nearly straight line that had not been there before.

Their return went smoothly at first, following their old tracks, but eventually the party lost its way in overcast weather and storms. Then rations ran short with the mountains still nowhere in view. By December 11 the men began to glimpse land during clearing moments, but they did not recognize where they were, and for the most part marched blind. On December 15 the ice surface began to descend, taking the party through an ice disturbance, still uncertain of location. The ice steepened and Lashly slipped, pulling the sledge and the other harnessed men with him. The group accelerated down the incline, bodies flying, as the smooth ice turned rough. The party was suddenly airborne, then stopped abruptly in a snowfield. The men were bruised, but they had no broken bones. Upon picking themselves up, they recognized where they were in the upper reaches of Ferrar Glacier, so set a course for Depot Nunatak and the food that they so urgently needed.

The day's adventures were not over, however, for as the three men descended the first of the two rolling icefalls, both Scott and Evans dropped into a crevasse. The sledge jumped the crevasse as Lashly braced and managed to hold it from slipping back into the slot, where the others were left dangling on their traces twelve feet down in midair, "with blue walls on either side and a very horrid-looking gulf below." As hoar crystals rained down on their heads, Scott, then Evans, managed to swing over to a narrow shelf of ice, where they stabilized themselves. With his one free hand Lashly managed to shore up the sledge with a couple of extra sled runners, but he was unable to let go lest it fall into the slot. Consequently, Scott had to shinny up the rope, a challenge almost beyond the strength that he could muster given his bulky clothing and rapidly freezing hands. But this crevasse was not to be his undoing, and Evans was helped in his ascent by some pull from Scott's harness that had been lowered to him.

That night in the calm lee of Depot Bluff, Scott's party savored warm hoosh and was relieved that the worst now lay behind. (Hoosh was a sort of thick soup made of crumbled biscuit and pemmican with a little water. Pemmican was the staple trail food, made of a combination of chopped and compressed dried beef, suet, and vegetables.) The descent of Ferrar Glacier would bring warmer temperatures and a series of replenishing depots. Having reached the Knobhead moraine by December 16, fit and well fed, Scott decided to follow the northern arm of the glacier that rounded the corner of the Kukri Hills. On December 17 the party pulled across Ferrar Glacier onto the northern arm. The ablation-pitted ice was rough on the sledge runners as the party descended. Several miles along, the glacier took a turn to the east. From there they followed a trail of boulders for several more miles down a sparse moraine, until the ice became too rough and the gradient too steep for the good of the sleds. At that point the party set up camp in the lee of

Figure 2.14. In the summer of 1903–1904, Scott and his party were the first to see one of the Dry Valleys to the west of McMurdo Sound. In this view to the northwest of central Taylor Valley, the deteriorating Taylor Glacier terminates at Lake Bonney. The groin of rock extending across the valley bottom restricts Lake Bonney to seventeen feet at the strait. The upper part of Taylor Valley is marked by dark and light layers of dolerite and sandstone. Scott's party hiked around the eastern end of the Kukri Hills (left rear of image), and eventually locked into one of the deep, longitudinal gullies that carry to the terminus. From the right, the descending icefall of the Rhone Glacier reaches almost to the tip of Taylor Glacier. Bound by fifty-foot walls of ice, the toe of the Hughes Glacier emerges at the lower left. In the right rear of the image, Finger Mountain is partially exposed on the left of the snowy gap to the horizon. The steplike tiers of icefalls taken onto the plateau by the parties of Armitage and Scott are vaguely visible in this gap.

a huge boulder where they dipped water directly out of a small meltwater stream whose soft gurgling gave pleasant harmony to their evening's retreat.

At 7:00 A.M. the next day, Scott, Evans, and Lashly were ready to go. They took a rope, ice axes, crampons, and some lunch in their pockets. This was a lightweight traverse, alpine-style. They left the sledge and tent and hiked down the glacier. Freed after so many days in their traces, the men would have been hard to keep up with. Looking down the glacier, they could not see where it ended, deducing that it dropped off somewhere ahead. Out maybe five miles farther a low ridge jutted from the right wall nearly crossing the valley floor. It gave the appearance of a groin or breakwater at the mouth of a harbor and certainly looked as if it would dam the glacier if it extended down that far (Fig. 2.14).

Fortunately, as the men hiked along, they found that the glacier did not have any crevasses, just sealed remnants that splotched the ice with whiter blues than the background. The surface was pitted with ablation cups, those knobby ripples so common on the hard ice surfaces of glaciers. They form by evaporation of the ice in patterns reflecting the angle of incident light as the circling sun rises and falls overhead in the summer sky. As the party moved along, the gradient became steeper and the ice began to show signs of wastage, similar to what they had seen at the mouth of Ferrar Glacier, though not on such a grand scale. Melting during the warmest days was surely active here. Patches of boulders and sand were sparsely strewn across an increasingly hummocky terrain. Partway along the party roped up.

A small stream channel appeared, starting in a shallow furrow parallel to the direction of glacier flow. Some discussion ensued about whether to stay high on the glacier or go down along the furrow. They chose the furrow with the rationale that the glacier might become too steep were they to follow it to its end, and that the channel might cut down to the base of the glacier over a gentler incline.

Figure 2.15. Taylor Glacier terminates in the permanent ice of Lake Bonney. Meltwater forms a thin moat around the shore of the lake and produces the deep blue ponds at the juncture with the glacier. Although the terminus is surely somewhat altered, one can imagine Scott's party traveling down the central groove in the glacier. The walls rise forty or fifty feet from the bottom. Where the gully terminates above the blue meltwater pond, there is a sheer drop of about twenty feet, but the point of ice to the left of the gully's mouth slopes down to the water level and might offer a possible route off the glacier. Blood Falls, the orange-colored cone protruding into the water at the right end of the glacier, forms where a subglacial stream debouches into the lake. The coloration is produced when iron-rich salt water from beneath the glacier contacts oxygen in the atmosphere and the iron oxidizes to the characteristic rusty colors.

The party worked its way down for two miles or so through an ever-deepening corridor of blue. The walls were littered with irregular blocks and pinnacles that now reached up twenty feet above their heads. The channel cut back and forth, but overall followed parallel to the glacier. Sand and gravel at the bottom of the channel made for fairly good traction. Then without warning the channel opened into air, and the men reached the end of the glacier, gingerly peering over an edge that dropped about twenty feet almost straight down (Fig. 2.15).

Taking a step back to behold the scene, the explorers witnessed an alien enclave in that realm of ice. The dark rocks of the valley before them were a sponge to the rays of the sun. It soothed their eyes. Clearly, a mighty glacier had once flowed down this smooth-bottomed valley and out to the sea, but now it was wasted and in retreat. At the toe of the glacier was a small lake fed by its melting waters. It was mostly frozen and whitish in color, but its edges were encircled by a narrow moat of liquid water. Immediately to the left was a gigantic hanging glacier that originated somewhere on the crest of the mountains and plunged at least three thousand feet down the steep, northern wall of the valley. Great blue seracs hung motionless throughout this graceful tongue, bound all around by fifty-foot vertical walls of ice.

Off to the right, the glacier appeared to offer a route to the valley floor. Carefully each man descended the steep ice, steadied by his ice axe, one cramponned foot at a time. The groin of rock that they had spotted from above appeared almost to block the lake about a mile out ahead, so they hiked over to see, following the rocky shoreline, which was an easier surface than the old, rough ice that covered the lake. As it turned out, the groin did not quite cross the valley bottom; instead, the lake at the toe of the glacier escaped through a narrow strait, only seventeen feet wide, and opened beyond to a larger

64

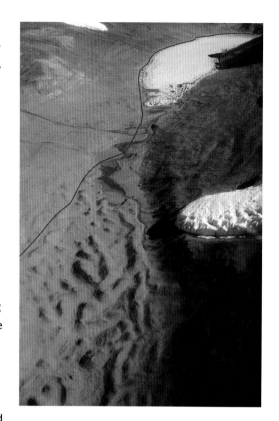

Figure 2.16. Hummocky, dissected, morainal deposits pattern the foreground, leading back to the eastern end of Lake Bonney. Scott's party hiked along the southern (left) side of the lake. He recorded that as they reached the end of the lake they quite suddenly "stepped out on to a long stretch, which here and there opened out into small, shallow lakes quite free of ice." This is the area crossed by the meandering stream that flows into Lake Bonney. It was on this fluvial deposit that the party lunched, dipped water from the stream, and felt the warmth of the dry sand draining through their fingers. Meltwater for this stream is coming from the toe of Matterhorn Glacier, on the right side of the picture, and a hanging glacier out of sight to the left.

lobe, three or four miles long. Scott was like a hound on the scent that day, striding forward around the lake. When everyone reached its end, he kept right on going along the sandy outwash that was being cut by a meandering stream feeding the frozen lake from the east (Fig. 2.16). Evans made an oblique query of Scott whether they needed to carry the lunches any farther, and the commander suddenly realized that it was well into the afternoon. They stopped for lunch: pemmican, chocolate, sugar, and biscuit, washed down with water dipped from the stream in their shared tin measuring cup. They sat in wonder on a bar of dark sand, sifting its grainy warmth through their bare hands, pondering the improbability of such a balmy pocket existing in this otherwise ice-bound land.

Scott declared that he would like to glimpse the sea through the lower end of the valley, so the party pushed on in that direction across a hummocky moraine field. When it became clear that the valley narrowed into a deep gorge farther to the east, the best possibility for a view seemed to be from the large shoulder that rose to the middle level of the valley farther on. It was a steep, long climb onto the shoulder (today known as Nussbaum Riegel), but once they reached its crest, the view down the valley was blocked by yet another shoulder five miles beyond. It was 4:00 P.M. They had been out for eight hours. Time now to turn back to camp. It would be an uphill slog all the way, but the party hit a cadenced rhythm and, ruminating on the wonders they had seen that day, made it back to camp by 10:00 P.M.

The next day, December 19, the party ascended the north arm, and by December 20 had reached the depot at Cathedral Rocks. The next five days were uneventful as they pulled back to *Discovery,* arriving on Christmas Eve. Only four men were there to greet the weary party. Again this season the sea ice had held fast, and all the rest of the men were out at the ice edge twenty miles to the north, where round-the-clock work groups were attempting to saw through the six- to seven-foot-thick ice to make a path for *Discovery*'s escape. After managing only 150 yards in twelve days, Scott called off the futile exercise.

On January 5, 1904, two ships, *Morning* and *Terra Nova,* sailed into view, bringing orders from England that if *Discovery* did not break free she should be abandoned rather than having the party winter over for another year. This was quite a blow to the men, who for two years had enjoyed the warmth and cheer of *Discovery*'s wooden walls. Consequently, a spirit of gloom rather than joy hung over the operation of transferring cargo to the relief ships.

By mid-January the ice had begun to break out a little, but by January 24 the front was still thirteen or fourteen miles away. On January 28 a long swell began to buckle the ice clear in to Hut Point. During the next two weeks, the front slowly broke up, and then on February 14 the whole of McMurdo Sound began to go. On February 16 *Discovery* finally broke free with the help of a blast of guncotton. Everyone was jubilant as the ship steamed out of Winter Quarters Bay. But a final blow awaited, for as the ship rounded Hut Point she struck rough seas and gale-force winds, which drove her aground. The next fifteen hours were nerve-wracking while the ship was pounded by waves and scraped against the rocks. At last the storm weakened and *Discovery* slipped into the water, free at last to return the expeditioners to their homeland and a hero's welcome.

The *Discovery* Expedition had added significantly to the map of Victoria Land (see Fig. 1.16). Scott and his men had charted the coastline south of Mount Melbourne in more detail and recognized the Drygalski Ice Tongue for what it was. They sighted the polar plateau through the cleft of Reeves Glacier and made a landing at Granite Harbour. The most important geographical discoveries occurred during the crossings of the Transantarctic Mountains via Ferrar Glacier. The expedition recognized that the mountains form a discrete range, that is, one with a backside, and that beyond the mountains the ice rises as a sheet across a vast plateau. They also mapped details of the mountains from the southern end of the Royal Society Range across Ferrar Glacier to Taylor Dry Valley, where Taylor Glacier had retreated, leaving an ice-free chasm the like of which had never before been seen.

3 Fire, Ice, and the Magnetic Pole
Further Discoveries in Victoria Land

One year before *Discovery*'s return, Ernest Shackleton had come home to his own hero's welcome, because news of the southern journey with Scott and Wilson had fired the public's imagination about hostile, polar wastelands bested by the explorers' will. Privately, Shackleton was deeply ashamed of his failure to hold up on the march, and, with the incentive of personal vindication, he sought to return to the Ross Sea and the icy interior beyond. He announced in 1907 his intention to lead an expedition to the Antarctic, which would attempt to reach both the Geographic and the Magnetic South Poles, not simply for the glory of England but also for sound scientific research. The party would be small, fifteen or so, composed of highly capable specialists. A single ship would leave the men on the continent and return to New Zealand for the winter, thus eliminating the need for a support vessel. He would use dogs to some extent for transport, but Shackleton, like Scott, had developed a mistrust of dogs during the *Discovery* Expedition; he planned to rely primarily on Manchurian ponies for hauling toward the poles. He also intended to use the first motorcar in Antarctica on the southern journey. This forerunner of modern-day snowmobiles was an open-seated contraption with large, wooden-spoked wheels, one of which is on display at the Canterbury Museum in Christchurch, New Zealand.

This was Shackleton's personal expedition. He worked without a committee, making all the decisions on the organization and outfitting himself. Without the financial support of the government or the societies, he was forced to make numerous appeals to wealthy individuals, mostly without success. Finally, enough commitments were forthcoming for him to begin seeking public support, but several promises for large sums were withdrawn, and in the end Shackleton assumed a £20,000 debt with repayment expected from the proceeds of lectures and a book upon completion of the expedition.

Following the publication of the plans for "A New British Antarctic Expedition," which included wintering at Winter Quarters Bay, Scott wrote to Shackleton politely but bluntly telling him not to use the *Discovery* hut. In fact, Scott had been quietly planning his own next expedition to the same area. Shackleton was stunned, but coolly revamped his carefully planned program, with a base as close to King Edward VII Land as possible, from which two parties would sledge in the region and a third would traverse southward across the Ross Ice Shelf to the pole.

The entire expedition was conceived in the most economical fashion. The best ship that could be afforded was a small forty-year-old Norwegian sealing vessel named *Nimrod*. The engines under full steam managed only six knots. The ship was delivered battered and reeking of seal oil on June 15, 1907. A speedy overhaul that included rerigging had the *Nimrod* ready for her departure from England on August 7.

At that time the party included two members from the *Discovery* Expedition, Frank Wild and Ernest Joyce, who assumed the responsibilities of the supplies, the general stores, and the dogs. The scientific team included James Murray, biologist; Raymond Priestley, geologist; Sir Philip Brocklehurst, assistant geologist also responsible for current observations; Lieutenant J. B. Adams, meteorologist; and Drs. A. F. Mackay and Eric Marshall, surgeons, with the latter also doubling as cartographer.

When the expedition arrived in Lyttleton, New Zealand, it received generous financial support from both the New Zealand and the Australian governments. At that time two Australians joined the party: Bertram Armytage for general work, and Douglas Mawson, a young geologist. The final man to join the group was fifty-year-old Professor Edgeworth David, a geologist from Sydney University, who according to the original plan would sail south and then return with the ship to Australia. David, however, was so universally liked by the members of the expedition that Shackleton prevailed on him to stay for the winter-over, thus significantly enhancing the scientific caliber of the venture. Imagine the reaction of David's wife when she received the letter that the good professor would not be making it home that winter.

To save coal and time, *Nimrod* was towed south by the *Koonya*, a 1,100-ton, steel-built steamer jointly financed by the New Zealand government and Sir James Mills, chairman of the Union Steamship Company. On New Year's Day 1908, a throng of thirty thousand crowded the Lyttleton docks to see the explorers off. Almost immediately the ships met stormy seas that lasted for ten days, soaking the passengers on the overloaded *Nimrod*. On January 15 the ships encountered their first ice floes. After a last exchange of mail, *Koonya* cast off and returned to Lyttleton, leaving *Nimrod* to her own resources.

As *Nimrod* pushed south, the ice remained in loose floes, not tightening into the dense pack that previous expeditions had encountered. By the next morning, the ice floes had given way to a vast company of tabular icebergs 80 to 150 feet high. Shackleton wrote,

All the morning we steamed in beautiful weather with a light northerly wind, through the lanes and streets of a wonderful snowy Venice. Tongue and pen fail in attempting to describe the magic of such a scene. As far as the eye could see from the crow's-nest of the *Nimrod,* the great, white, wall-sided bergs stretched east, west and south, making a striking contrast with the lanes of blue-black water between them. A stillness, weird and uncanny, seemed to have fallen upon everything when we entered the silent water streets of this vast unpeopled white city.

In hindsight, a rare, major breakout from the Ross Ice Shelf probably produced these conditions, with the bergs disrupting the pack as they were blown through it by southerly storms.

By 3:00 P.M. on January 16, *Nimrod* was in open water, so she set a course southeast. On the morning of January 23, "a low, straight line appeared ahead of the ship": the Ross Ice Shelf. The plan was to put in at a low point at Balloon Bight and build a hut for a base on the ice shelf. But as the ship approached the location of what Shackleton remembered to have been a narrow inlet in the ice shelf, he found a broad bay encompassing both Balloon Bight and the inlet where Borchgrevink's party had landed, seven or eight miles to the east. Because of the numerous sightings that day, Shackleton named this new feature the Bay of Whales.

The fresh face of the ice shelf was too high to ascend, so Shackleton shifted to his backup plan of establishing a camp in King Edward VII Land. A closing pack, however, prevented the ship from approaching that land within fifty miles, so Shackleton had no choice, if any effective base were to be established, but to sail back to McMurdo Sound.

With expectations of reaching Winter Quarters Bay, the *Nimrod* was stopped twenty miles short by fast ice on January 29. After a three-day sledge trip to Hut Point, which found the *Discovery* hut well provisioned, Shackleton decided to build his winter quarters at Cape Royds. Thereafter began an intense two-week period of unloading the ship, during which all the men worked around the clock, catching a few hours' sleep only when they no longer could continue. By February 18 almost all stores were safely on shore, and the hut was built. After a severe blizzard that drove the *Nimrod* offshore for three days, the ship was emptied of a last load of coal and steamed back to the warmth of New Zealand for the winter.

Cape Royds was a happy choice for a base, because it is the location of an established Adélie penguin rookery (Fig. 3.1). The hut was situated beside a small frozen lake across from the rookery in a hollow that practically obscured the view of the mountains across McMurdo Sound but appeared to offer shelter from the wind. The steam-plumed summit of Mount Erebus towered fifteen miles behind the camp (Fig. 3.2). In contrast to the *Discovery* hut (which was at the end of a peninsula twenty-five miles distant from the Erebus summit and without a view unless one climbed Observation Hill), the hut at Cape Royds was on the flank of the volcano with the summit looming behind.

As the final details of establishing the base were completed, discussion picked up on what to do next: What would be the best, first goals of the expedition, and the scientific endeavors? With an active volcano barely five leagues away and four geologists in camp after taking on the Australians, there was a strong argument for investigating the geological phenomenon just up the hill.

The ascent of Mount Erebus became the first sledging venture of Shackleton's expedition, including an advance and a support party consisting of three men each who would

Figure 3.1. Adélie penguins couple on the rookery at Cape Royds about one hundred yards from the doorstep of Shackleton's hut. A group of Emperor penguins huddles next to the ice front in the middle ground, while several Weddell seals lie on the ice to the right. Mount Discovery rises in the center distance to the southwest across McMurdo Sound.

Figure 3.2. Rising to 12,450 feet, Mount Erebus steams behind Shackleton's hut at Cape Royds. David's ascent route to the summit is shown in orange.

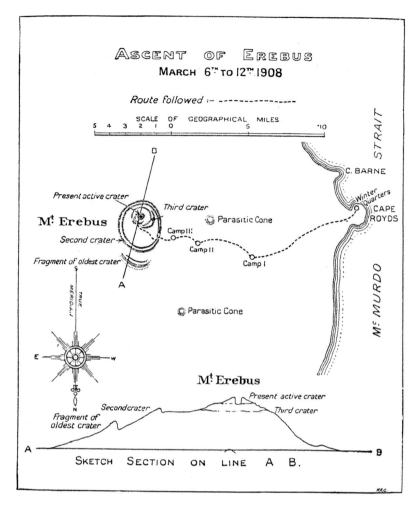

Figure 3.3. David's map of the ascent route to the crater of Mount Erebus may be matched with Fig. 3.2. The lower two spots on Fig. 3.2 are Camps II and III. The uppermost spot, where the party camped at the edge of the "Second Crater," is not shown on David's map.

haul an eleven-foot sledge. The advance party would have crampons, one-man sleeping bags, and the means to backpack with a tent and stove to higher elevations. The support party, with no crampons and a three-man sleeping bag, was not (in Shackleton's mind) expected to summit, although he did allow the leaders of the parties the authority to decide on the mountain whether the support party would attempt the crater.

The advance party consisted of David (leader), Mawson, and Mackay (two geologists and a physician), and the support party, Adams (overall leader), Brocklehurst, and Marshall (a meteorologist, a geologist, and a physician). They set out together on March 5. Once on the trail, none of the men would be denied. After two days of pulling the sledge over all manner of ice and snow as well as carrying it over rocks, the parties camped at an elevation of 5,630 feet at minus 28° F (Camp II, Fig. 3.3). To the man they wanted to creep to the edge of the volcano and peer into its wheezing maw. By then, they were a team in the trusses, and Adams gave the nod, even though the support team would be hampered by its large sleeping bag and at a disadvantage on steeper ice without crampons.

The next day the men deposited the sledge and some supplies at a depot and set off, each carrying about forty pounds on his back. The support party had improvised pack boards, but without crampons they had to cut steps when they encountered the steeper ice—a tedious, tiring job of chopping with the ice axe, bent over, trying to keep balance in two aligned holes in the ice while chopping the next. David had given Marshall his crampons and used a set of leather strips on his ski boots that he tested to good success, but Adams and Brocklehurst struggled over icy stretches of the ascent.

The party carried two tents but had left their tent poles at the depot. They camped at 8,750 feet in their sleeping bags inside the limp shell of their tents (Camp III, see Fig. 3.3). During the night a blizzard sprang up and pinned them down for another full day, during which they had no water because they could not use the cooker in the tents. Imagine being in Adams's tent, one of three men in one big sleeping bag: every time someone moved, so must everyone else. At one point, Brocklehurst and Adams had to go outside. They managed to position the sleeping bag to squirm out of the tent. As Brocklehurst was taking off a mitten, the wind grabbed it from his hand. As he tried to rescue it, the wind took him too, and sent him tumbling down a ravine. Adams jumped to catch Brocklehurst and was also blown down the ravine. Marshall was in the awkward position of losing the sleeping bag to the wind if he tried to rescue his mates, so he held tight. Both Adams and Brocklehurst made it back to the tent by crawling on their hands and knees, and everyone was fine, but it had been a close call.

On the morning of March 9, the climbing was the steepest thus far, with a slope of about 35°. The men were able to piece together some rocky stretches, but wherever there was ice they had to cut steps. They set up a lunch camp in a rocky gully about fifty feet below the edge of an old crater rim. Brocklehurst admitted that he had not had feeling in his feet for quite some time. When Marshall removed Brocklehurst's boots, the men were shocked to find that two of his toes were black, indicating that they had been frozen for a number of hours, and four other toes showed lesser signs of frostbite. Marshall and Mackay warmed up his feet, gave him dry socks and an extra cut of sennegrass for his finnesko (reindeer leather boots with hair on the outside), and tucked him into the three-man bag to rest while the others spent the remainder of the afternoon exploring.

The inner side of the old crater was a steep-to-overhung wall eighty to one hundred feet high that dropped into a wind scour about thirty feet deep. Beyond, a fairly smooth névé climbed to the summit of Erebus. The party found a cleft in the crater wall, where a drift of snow had bridged the wind scour, and walked the line across to the névé. They were drawn immediately to the bizarre ice mounds scattered across the névé that they had spotted from the rim. It was like the men were in some postapocalyptic sculpture garden. The forms, all rounded and knobby, stood tens of feet at their highest (Fig. 3.4). They resembled turrets and chimneys, beehives and haystacks; some were grotesque animals with gnarly sinews ready to pounce. Upon closer examination they found the mounds to be largely hollow. Professor David concluded that these remarkable structures were forming at the vents of fumaroles, where steam escaped from the volcano. (Fumaroles are common on most active volcanoes, but the difference here was the subzero temperatures, which caused condensation and freezing of the moist vapors as they issued from the mountain. What a great place to take a sauna—if you could stand the sulfur fumes!)

The party climbed about a mile up the névé to a small parasitic cone with good exposure. The geologists were delighted with what they found. The groundmass of the rock

Figure 3.4. Ice fumaroles stand as sentries near the summit of Mount Erebus. Formed by the freezing of vapors emitted by the volcano, the taller structures are about ten feet in height. Kenyite lava crops out on the right.

was pumice, a frothy, brown glass that crumbled with the force of their thumbs, chock full of the finest faceted crystals of a rare type of feldspar called anorthoclase. The lavas around Cape Royds also were full of anorthoclase crystals, but there they were not much more than an inch in length and stuck solidly in the rock. Here the crystals ranged up to three inches, some having the classic rhombic shape, others with intricate twins protruding as symmetrical adornments on the primary crystals. Because the pumice was so crumbly, the anorthoclase weathered out of the glass, and was so concentrated at places that it almost formed scree. Most of the crystals were coated with a pale yellow crust of sulfur that added a burnish to their otherwise gray coloration.

Professor David was the one who had first identified the anorthoclase back at the winter quarters. He explained that they were characteristic of a volcanic rock called kenyite, named for its best known occurrence, in East Africa at Mount Kilimanjaro and Mount Kenya. What Mount Erebus shares with its East African kin is location—not geographical location, for they are at the extremes of latitude, but rather geological location, with each being found in the interior of a continent. More typically, volcanoes form in arcs at continental margins or in ocean basins, where one would never find kenyite. Everyone picked up a pocketful of the crystals before heading back to camp. They would make wondrous gifts for the grandchildren, when the old men told their tales of fire and ice.

The day was mercifully calm after the previous day's storm. The men easily lit the stove outside the tent and brewed tea. It was a pleasant meal, as they looked out to the west across a vast layer of cumulus cloud banked halfway down the mountain and spread far across McMurdo Sound. The western mountains shimmered in the glow of the setting sun.

Up the next morning at 4:00 A.M., everyone but Brocklehurst set out for the summit about two and a half miles away. They marched past the ice fumaroles standing sentry to the fortress. From Cape Royds the plume had been like a siren's veil wafting gracefully from afar. The challenge had been in the approach across the fields of ice and stone. But now they were drawing close. The plume was no silken wisp, no insubstantial scarf. It had loomed ever more powerfully as they had risen toward its hidden source. From camp that morning as the cloud billowed broadly above the top, it felt as though they were approaching a dragon's lair. But now on the last steep pitch, as they scrambled breathlessly on scree of pure crystal, deep, ancestral fears swelled in their chests. This was surely the home of a god. Vulcan, Pele, Agni, different folk had used different names. Mostly these gods slept, wheezing and belching foul breath, but if awakened they ignited the havoc of molten rock, fire, and explosion. Another fifty feet to the edge. The steam was rising as a billowy wall straight out of the ground. They could hear the constant hiss as it issued from somewhere far below. As the men drew close, they sensed the mountain give a deep shudder every several minutes, more a feeling in the body than a sound. But now as they approached the rim, the feeling was audible, a deep boom from within the mountain, followed by a surge of roiling steam. The anxious mortals crept to the edge and peered in, praying that the god would not be awakened.

Nothing but murkiness, the rising swirl of steam, the stench of sulfur fumes. Why do we have such fascination with these lesions on Mother Earth? Is it the power revealed, the instinct for sacrifice and purification, the chance to witness the spirit of a possessed mountain? The men trembled in expectation.

Then quite suddenly a breeze from the north lifted the veil from the bottom, and there was the crater in its entirety, a nearly perfect cylinder a half-mile across, dropping nine hundred feet vertically down to a flat floor (Fig. 3.5). They could make out three centers where most of the steam was venting, circular holes reverberating with the shrill release of pressured gas, and one of them was the source of the periodic boom. Alternating horizons of black and white patterned the far wall. A whole row of tiny steam jets rose from the top of the darkest pumice layer, coating the crater wall in a veneer of icy hoar (Fig. 3.6). What the men were not able to see because of the steam was the floor of the inner crater, where a lake of lava continuously churned with convection currents (Fig. 3.7).

The geologists worked busily—so much for worrying about waking the god! Mawson measured the crater dimensions and took photos. David collected fresh pumice and sulfur. Both sketched. They noted volcanic bombs nearly a foot in diameter, hunks of pumice coated with sulfur thrown out on the rim. Lava must have risen close to the vent at times and been coughed out like phlegm from the volcano's throat, but today it was less active. Each man took one last look at this natural wonder, listened to the hissing, whiffed the sulfur, pocketed a few more crystals, and then headed back. It was time to be getting down the mountain. By 3:00 P.M. the party had a good meal at camp, and then the men shouldered their packs. Brocklehurst bravely insisted that he carry his own

Figure 3.5. The summit crater of Mount Erebus is indicated in Fig. 3.3 as "Present active crater." David measured the wall from the upper edge to the floor as nine hundred feet in height. An inner crater drops into the floor.

Figure 3.6. As described by Professor David, steam rises from the dark layer on the inner wall of the summit crater of Mount Erebus.

Figure 3.7. Unseen by David's party during the first ascent of Mount Erebus, a lava lake that continuously convects within the inner crater.

despite his frostbite. With the goal of descending as fast as possible, the party glissaded whenever it reached the snow stretches free of rock.

To perform a glissade, you slide on your feet as though they are skis, using the tip of your ice axe as a brake. Grasp the top of the ice axe in one hand and with the other grab about halfway down the shaft. Now hold it so that the arm on the axe head is across your body and the hand holding the shaft is against the side of your hip, with the point sticking down and behind you. Bend into a half-crouch, as if you are sitting in a chair, and put weight back onto the point of the axe. Lean forward, taking pressure off the ice axe, and pick up speed; put weight back onto the axe and slow down. If you get going too fast and fall, be sure to roll over on your stomach and do a self-arrest with the point of the axe head; otherwise, you could build up too much speed and wipe out.

The one hundred–centimeter axe that the men used was perfect for this maneuver. It also doubled as a walking stick on rough terrain, saved them from bending over too far when cutting ice steps, and could be used as a handy probe for crevasses. In the late twentieth century, when ice climbing became all the rage, the alpinists traded in their old-style axes for ice "tools," a pair of stubby picks that could be used for climbing frozen waterfalls. Glissading became a lost art, and everyone stumbled a lot more than they once did with the old reliable long axe providing balance.

By 10:00 P.M., the men had descended five thousand feet and reached the depot they had created on March 7. Everyone was pretty bruised from all the spills that they had taken, the sastrugi (windswept surface patterns on snow) having been rough in places, but now they were breathing thicker air again, and the hoosh warmed their bellies.

The party was on the trail by 5:30 A.M. the following day; by 7:30 A.M. it had reached the depot of March 5. With storm clouds threatening and the hut in sight only seven miles away, the men decided to cache their sledge and gear and make a dash for it. Although the morning was overcast so that definition on the sastrugi was lost, the blizzard never quite descended, and everyone was back by 11:00 A.M. Shackleton broke out the champagne to toast the returning conquerors, who filled the hut with stories of their exploits.

When it came to food, Shackleton relates,

> Except to Joyce, Wild and myself, who had seen similar things on the former expedition, the eating and drinking capacity of the returned party was a matter of astonishment. In a few minutes Roberts had produced a great saucepan of Quaker oats and milk, the contents of which disappeared in a moment, to be followed by the greater part of a fresh-cut ham and home-made bread, with New Zealand fresh butter. The six had evidently found on the slopes of Erebus six fully developed, polar sledging appetites. The meal at last ended, came more talk, smokes and then bed for the weary travelers.

Winter passed favorably for the fifteen men at Cape Royds. Everyone was active enough with chores and pastimes. The dogs and ponies needed regular tending. Birthdays and holidays were celebrated without fail. Wild and Joyce produced a 120-page book, *Aurora Australis,* and printed one hundred copies on a press donated to the expedition and bound them with board from packing cases.

Every two hours the expedition made a meteorological observation, which required someone to go outside. In starlight or moonlight the plume on Mount Erebus was always noted for determining the wind direction at twelve thousand feet. The progress of activity at the summit was a focus of continuing interest. On numerous occasions, a red glow from the crater was reflected on the bottom of the plume, and at times someone would report "great bursts of flame crowning the crater."

They planned two major traverses for the following summer. The southern party, led by Shackleton, would attempt to reach the South Pole, and the northern party the Magnetic South Pole. Spring was filled with a number of sledging trips designed to provide experience to the uninitiated for the more sustained efforts later on. These trips mainly involved laying depots for the southern party.

The northern party consisted of Professor David (leader), Mawson, and Mackay. Their route would cross McMurdo Sound to Butter Point and proceed northward from there along the coast, on sea ice if possible, on the piedmont glaciers if not, to a point where they could gain access to the plateau through some gap in the mountains (Fig. 3.8).

Except for the motorcar, which successfully hauled loads on sea ice around winter quarters but was of no use on rougher terrain, the party had no support. On September 20 the men hauled one of the two sledges the party would be using ten miles out onto the sea ice. The motorcar pulled the second sledge, and the three-man party began its odyssey on October 5 from Cape Royds. The vehicle had driven only two miles when thickening snow forced it to turn around, leaving the men to haul for the remainder of the day in mixed conditions, reaching their previously deposited sledge at 7:00 P.M.

Because the combined weight of the two sledges was too great for the party to pull as

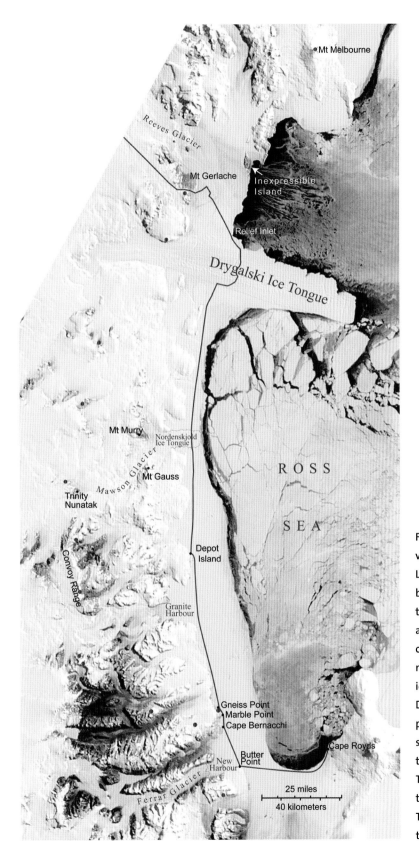

Figure 3.8. This satellite view of coastal Victoria Land traces the route taken by David's party from Winter Quarters at Cape Royds across the southern margin of McMurdo Sound and northward along the fast ice margin. North of the Drygalski Ice Tongue the party turned inland and struck a course straight to the Magnetic South Pole. The magenta dots show the survey stations of the Topo North party (Operation Deep Freeze 62).

a pair, the men were resigned to relaying them for many weeks to come. First, they would haul the sledge with camp gear and scientific instruments for a third or half a mile, then they would hike back to the sledge loaded with provisions and haul it up to the first. This procedure meant that on the outward leg of the journey they would be covering their distances three times.

On the morning of October 13, they reached Butter Point, having taken eight days to travel a distance that some previous parties had achieved in two. Already they were well behind the schedule Shackleton had set. To lighten their loads, they cached seventy pounds of food and gear at Butter Point and turned north, crossing New Harbour the next day. On October 15, one of the finest days of the journey, the men enjoyed excellent views up Ferrar Glacier. David wrote, "Towards evening we had a wonderful vision of several large icebergs close ahead of us; it seemed that they were only a mile or so distant, as one could see clearly the re-entering angles and bright reflected sides of the bergs lit up in rays of the setting sun. Suddenly, as if by magic, they all vanished. They had momentarily been conjured up to our view by a wonderful mirage."

For the latter half of October, the party worked its way north along the coast, averaging four or five miles on a good day. Conditions on the sea ice varied greatly. At places the ice was mercifully smooth; in others, it was cracked and shingled by compression. On several occasions, the party had to cross leads over thin, recently frozen ice. Rough sastrugi were common obstacles, and freshly fallen snow always was a drag on the hauling. Back at Cape Royds, Mackay had rigged a sail out of the tent bottom and poles, but it could be used during precious few intervals.

The view of the mountains north of New Harbour was attenuated across a broad, ice-covered piedmont. Although high ranges appeared to rise behind the foothills, the character of these mountains remained vague. Scattered along the shore of the piedmont were a number of promontories and small islands, which the geologists visited and sampled at every opportunity. The buffet offered a variety of metamorphic rocks and granite. At each camp Mawson would shoot a round of summits with his theodolite (a surveying instrument with a telescope that measures vertical and horizontal angles), mapping the positions of the mountains as the party moved northward. On October 25–26 the party passed the mouth of Granite Harbour, resisting the temptation to "geologize" among the cliffs that Scott, Wilson, Shackleton and Koettlitz had spot-sampled seven years before. Scott's original plotting of Granite Harbour from the *Discovery* had it twenty miles farther north than David's party found it to be, based on the readings from their sledge meter (a wheeled device attached to the back of a sledge that recorded distance traveled).

The rate at which the party moved was woefully inadequate to allow any chance of reaching the magnetic pole. The only hope for increasing speed was to lighten the load. As early as October 23, David had proposed the possibility of caching food and supplies and marching forward on half rations. Mawson and Mackay had agreed. For the next week, they mulled over the details and kept a lookout for a proper spot for laying their depot. On October 30 they picked the spot, an islet three-quarters of a mile offshore and ten miles north of the mouth of Granite Harbour. For the geologists this prominent point of land, which they called Depot Island, had the added attraction of a suite of exceptionally large and unusual metamorphic minerals.

THE SILENCE

During my first Antarctic field season (1970–1971), I worked from a remote field camp on McGregor Glacier, a tributary to Shackleton Glacier midway along its east side. I arrived on a ski-fitted Hercules C-130 aircraft that set down on McGregor Glacier. When I got off the plane, I was ending a long series of flights that had begun three weeks earlier in Quonset Point, Rhode Island, on the other side of the world.

Three navy helicopters arrived the next day. They would serve the camp, tasked with putting out field parties in the morning and then picking them up at the end of the day to bring us back to base camp, where we would sleep and eat.

Two days after I arrived I was flying out to my first day in the field. My excitement was acute. I had picked a site about eight miles north of camp, a small island of rock on the eastern side of Shackleton Glacier named Taylor Nunatak. The helicopter dropped Phil Colbert and me, along with our survival gear, and flew away. Its rotor faded and then was no more.

For the first time in weeks, I was without the audible vibe of engines, whether screaming from aircraft or just humming to themselves somewhere in the background, quietly unnoticed. It wasn't apparent at first, for as Phil and I scurried over to the margin of the glacier, we generated sounds of our own— footsteps, the click of our ice axes, the swish of our clothes. We stopped on a terrace of rock and looked out at Shackleton Glacier across a spectacular icefall that gave the appearance of rapids in a wild river, except that it was motionless.

Standing there, I suddenly became aware of the silence. It was behind me just at my shoulder. It went beyond the icefall as far as I could see. It was out there everywhere. The stillness was profound. The soft rustling of my parka seemed amplified. I held stone still. The sound of breath issuing through my nostrils filled my ears. I held my breath. The silence pressed in on all sides. It was palpable (Fig. S.6).

In my state of auditory suspension the icefall before me was all the more dramatic. It occurred at a point where a small ridge of rock projected from Taylor Nunatak into Shackleton Glacier. The glacier margin first reared up over this obstacle, broke into a jumble of blue seracs (ice blocks), then plunged chaotically down a steep, two hundred–foot scoop at the margin of the glacier, before molding smoothly back into the flow. In *human time* the icefall was a static sculpture of Nature's grand design, in *glacier time* (moving about five feet per day) it was a rapids in a smoothly flowing river.

As Phil and I stood there savoring our solitude, a sudden thwack broke the silence, like the shot of a .22-caliber rifle. We jumped and looked around, but there was no apparent source of the sound. A couple of minutes later another shot rang out. This time we were sure that it had come from the direction of the

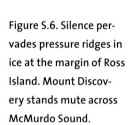

Figure S.6. Silence pervades pressure ridges in ice at the margin of Ross Island. Mount Discovery stands mute across McMurdo Sound.

icefall. Then it happened again. The ice was popping, strained to its limit and fracturing, incrementally working its way down the gradient at less than a millimeter per pop. The icefall defied the silence. The dead air sucked up its sound.

At last I was really out there, out there in a state of extreme isolation, even grace, with raw Nature everywhere on display. It was time to see what secrets the rocks would tell, and we went clomping up the outcrop, challenging the silence at every step.

Because they would be traveling on half rations to the point of turning inland, it would be necessary to supplement their diet with seal and penguin. Conserving paraffin for their primus stove was also an issue. The inventive Mackay devised a stove out of a cookie tin for burning rendered seal blubber. The bottom of the tin held lumps of blubber that were heated from below by the primus. He devised a wick holder from the lid of the tin by punching a series of holes for the wicks and bending down legs from the corners to make a little table that was set in the bottom. As the water from the fat sputtered off and pure oil ran down, wetting the bottoms of the wicks, the stove burned warmly without further help from the primus.

As the weeks had passed and the sea ice had increasingly broken up, the men knew that their escape route was disintegrating behind them. Coupled with the doubtful prospect of man-hauling back over a mountain route, they were now committing their survival to *Nimrod* and her ability to find them along this desolate coastline. Mackay built a prominent cairn on the top of the island with a pole firmly planted in its core and a black flag topmost. He cached geological specimens by the cairn, and wired a dried milk tin

Figure 3.9. In this image encompassing a stretch of coastline followed by David's party to the Magnetic South Pole, open water and very loose pack extend back to fast seasonal ice connected to piedmont glaciers that rim low tabular mountains behind. The skirt of seasonal ice was what David's party followed for most of their traverse. Nordenskjold Ice Tongue issues from the range a little left of center. In David's time the ice tongue was estimated at twenty miles in length, much longer than the calved stub of today. The extended length was what made the party choose to sledge across it. The portals of the Nordenskjold Ice Tongue are Mount Murray to the north (right) and Mount Gauss to the south (left).

to the flagstaff as a postbox. Among the letters left was one with specific instructions to the *Nimrod* to look for a similar cairn along the "low sloping shore" to the north of the Drygalski Ice Tongue, where the party intended to turn inland for the magnetic pole and hoped to rendezvous at the end of their trek. Letters to loved ones were also included in case the explorers did not make it back alive.

On the morning of November 2, after placing the last of their letters in the postbox, the men started north, soon encountering the hardest pulling of the trip. The sun was thawing the brackish snow surface, which stuck like glue to the sledge runners, so they decided to change to hauling at night, in hopes that the colder temperatures would keep the snow frozen. This adjustment helped in part, but fresh snow squalls continued to slow the party.

By November 11 the party had traversed thirty miles from Depot Island to the southern edge of the Nordenskjold Ice Tongue (Fig. 3.9). Charting of this feature by the *Discovery* Expedition indicated that it extended into the Ross Sea for a distance of twenty miles. To avoid taking a long detour to the east, David hoped to be able to haul the sledges directly across it. The ascent from the sea ice to the ice tongue was fairly easy, and the surface of the ice tongue itself provided some of the easiest pulling of the journey. On an exceptionally clear day from a camp in the middle of the ice tongue, Mawson was able to triangulate on Mount Erebus and Mount Melbourne. The mountains to the west were quite subdued compared with the high peaks that rose along the coastline north of Mount Melbourne, or compared with the mighty ramparts along the central Ferrar Glacier. The confines on either side of the glacier feeding Nordenskjold Ice Tongue (later named Mawson Glacier) were only about three thousand feet in elevation, with Mount

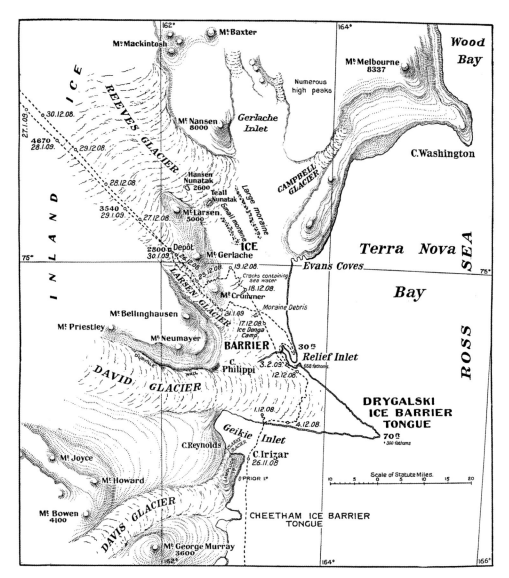

Figure 3.10. Mawson's map of the Terra Nova Bay area shows the route taken by David's party along the coast and up onto the plateau en route to the Magnetic South Pole. Note the pointed aspect of the Drygalski Ice Tongue in 1908, compared with the square end of recent years (See Fig. 1.13).

Gauss a little more than four thousand feet. Behind this front range the explorers could make out several low-relief tiers of mountains rising back to the polar plateau, but the highest barely stood out against the ice.

Crossing between the Nordenskjold and Drygalski Ice Tongues over varied sea ice took a little more than a fortnight. The men had decided an end run of the Drygalski would be too much time lost, so they attempted a direct crossing. The first try failed (December 1, 1908, Fig. 3.10). The party managed only a half-mile the first day onto an increasingly hummocky surface of blue ice, with patches of high sastrugi and chasms that dropped deep into the ice tongue. After a reconnaissance on foot, they decided that it would be useless to attempt a crossing, so they retreated back to the sea ice and began to pull toward the end of the tongue. A valley into the glacier tongue soon opened, however, and after a day of reconnaissance, they chose it as a possible route across.

On December 6 a hard pull up the steep, rolling snow surface put the party in a straight, snow-filled valley that ran for about two miles across the glacier tongue. But beyond that, the surface became hummocky blue ice again, with numerous crevasses and steep rolls and ridges up to forty feet high in places. The men managed to piece together a route that zigzagged into the field. The next day they found conditions to be just as bad. David described what he saw while searching for a route: "The surface still bristled with huge ice undulations as far as the eye could reach. It was just as though a stormy sea had been frozen solid, with the troughs between the large waves here and there partly filled with snow, while the crests of the waves were raised by hard ridges of drift snow, terminating in overhanging cliffs, facing north."

As the party neared the edge of the ice tongue, smoother snow conditions mirrored those on its south side. By the morning of December 11, the men had finally descended from the Drygalski Ice Tongue onto the fast ice that bordered the shore. That evening they camped next to a high mound of boulder-mantled ice, which the three climbed after dinner to take in the scene.

Consider the view they had—a magnificent scene it was! The coastline arched gracefully out to the north-northeast about sixty miles to Mount Melbourne, the perfect triangle that stood prominently on the horizon (Fig. 3.11). The thin line that ran out from it on the right was Cape Washington. Off to the west of Mount Melbourne around the deepest part of the reentrant were some formidable mountains. They appeared to be peaked and cut deeply by valleys. Maybe Scott didn't have a good view up to the northwest when he steamed through on the *Discovery*. Regardless, the mountains to the north were beautiful to behold. They might be seven thousand or eight thousand feet in altitude. What would they be like close up?

From those mountains, a steep escarpment ran down the coast, bare of snow in its upper slopes. The last buttress on this cliff (Mount Nansen) stood more than eight thousand feet (see Fig. 3.11). Then came Reeves Glacier, with its tiers of icefalls spilling from the mute plateau. The dark, beehive-shaped nunatak in the middle of the glacier was the one noted as a landmark by Scott in the summer of 1901–1902. Left of that was a broad, flat-topped buttress, with the highest point at the northern end named Mount Larsen. This side of that was another, smaller glacier (Larsen Glacier), and then a continuation of the buttress, which at its termination banked the vast glacier feeding the Drygalski Ice Tongue (later named David Glacier).

All the way from Mount Melbourne to within about a mile north of this vantage were the inky black waters of Terra Nova Bay. The party appeared to be in a section of ice that was fast to the shore in a sort of pocket between the Drygalski Ice Tongue and some bare bedrock islands about halfway up the coast (Inexpressible Island). It was hard to tell whether that ice was from the glaciers spilling over the mountains to the northwest or was sea ice that has survived many seasons since the last deep breakout.

On the Admiralty chart the stretch to the north of Drygalski Ice Tongue was indicated as "low sloping shore." But there did not appear to be a "low sloping shore"—only the old ice at the near end of the bay. This could be a problem when *Nimrod* came looking for the party at the beginning of February.

The conversation among the party members was of possible routes onto the plateau.

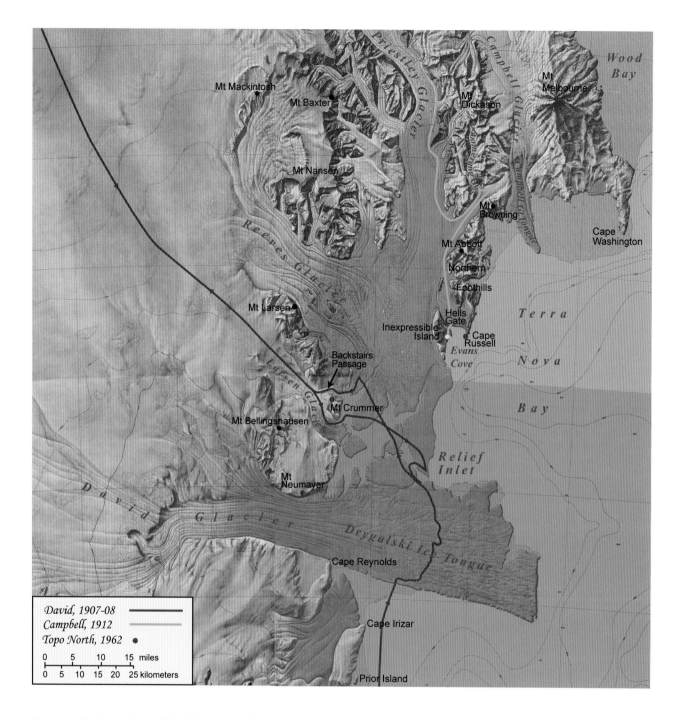

Figure 3.11. Cartographers of the U.S. Geological Survey used aerial photography taken in December 1963 to draw this shaded-relief map of Terra Nova Bay. The details of the sea ice around Relief Inlet and the outline of the Drygalski Ice Tongue had probably changed appreciably since the early twentieth century. In 1901 there was a deep reentrant in the area of Relief Inlet that *Discovery* nosed into before sailing back along the margin of the ice tongue. For David's party, which named the place, there was a distinct cleft in the fast ice, a portion of which was mantled by thick moraine. David's route for this figure follows the basic shape of the route on Mawson's map (compare Fig. 3.10) and places the end of the journey close to the edge of fast ice as indicated in 1963. A further comparison can be made of differences in the shape of the Drygalski Ice Tongue as mapped by Mawson (Fig. 3.10), as mapped by the U.S.G.S. (this figure), and as photographed in November 1986 (Fig. 1.13).

What thoughts they had had of dragging farther north over sea ice before turning inland were now out of the question. All the previous season's ice had left Terra Nova Bay. As a possible route the Larsen Glacier seemed fairly steep in its lower reaches. Although the crevasses on Reeves Glacier looked as if they could be horrendously challenging, the incline looked less steep, and maybe the crevasses would be mercifully bridged and solid. Clearly, the party was not going back onto the Drygalski Ice Tongue.

Although the mound on which the men were standing was about a mile from open water, it was by far the most prominent landform close to the ice edge in the area. There they erected another flagpole and cairn, and placed in a depot all but the minimum that they determined they would need for seven weeks on the trail, including one of their sledges. Mawson estimated that they were 220 miles out from the pole. At 440 miles out and back and forty-two days of rations, the men had to march a little more than ten miles per day, every day. If they were slower, they would miss the appointed rendezvous with *Nimrod*. To cut back on cooking fuel, as well as save cooking time in the tent, Mackay spent a full day cooking seal and penguin meat, while David and Mawson sorted the camp into what items were to go with them and what would remain behind. Finally, they wrote another round of letters to loved ones and comrades, in case the party did not make it back to the depot.

On December 16 the members of the northern party emerged from their tents after a two-day blizzard and hauled off to the west over a twenty-mile stretch of old, fast ice and distal glacier. The surface was melting wherever rocky debris was present, creating slushy bogs, shallow pools, and streams that flowed both on the surface and within the ice. Crevassing became increasingly dangerous. One swallowed Mawson to the length of his harness rope, and only with great difficulty did his mates extract him.

On December 20 a blizzard that dumped six inches of snow stopped them, and with the base of the Reeves Glacier still several miles off, the men decided to pull back and try Larsen Glacier. But as they drew near, the base of that glacier also looked too steep and badly crevassed. Leaving their sledge, the men hiked across to the base of the Mount Larsen block, where they found a small glacier that poured over the rock buttress, apparently as an offshoot of the main Larsen Glacier (Figure 3.12). Its surface was covered by soft, freshly drifted, knee-deep snow, but it appeared to be crevasse free.

Before the party could begin its ascent, another blizzard came down, with higher temperatures and much melting inside the tent. By the afternoon of the second day, the men were so miserable that they went out in the gale and dragged their sledge and half a load up to the eight hundred–foot elevation, deposited the load in a depot, and returned to camp. The next day, when the blizzard had blown itself out, they were pleased to find that the wind had also removed most of the snow from the previous storm, and now the surface was relatively hard snow that made for easier pulling. By 10:00 P.M. on December 24, the men were at twelve hundred feet and above the level of melting. On Christmas Day they arrived on the main body of Larsen Glacier and camped at two thousand feet. The name they gave to the ramp that connected them to upper Larsen Glacier was Backstairs Passage.

Two days later, when it appeared that the party had finally reached the plateau, the

Figure 3.12. A polynya (area of open seawater surrounded by ice) at the mouth of Reeves Glacier remains open due to the katabatic winds that pour down from the plateau and over Inexpressible Island into the southern end of Terra Nova Bay. David's party made its way into the left (south) side of the picture along old sea ice after crossing the Drygalski Ice Tongue. Its route is shown in red. As the party headed toward Reeves Glacier, with the two prominent nunataks in midstream, the men encountered impassable crevassing. When they turned back, they found a small overflow glacier (Backstairs Passage) to be a route onto the upper Larsen Glacier and the plateau beyond. Campbell's party was dropped by the *Terra Nova* on Inexpressible Island at the yellow dot. From there the men worked to the north (right) of the image, around the mouth of Priestley Glacier, before being forced to winter over at the beach where they had been dropped. Cape Russell, occupied by the Topo North survey party in late December 1961, is the thin peninsula of land in front and to the right of the yellow dot on Inexpressible Island.

men made a small depot of their geological specimens, the ice axes, climbing ropes, and ski boots, plus a day's worth of food. From there they set a straight course to the northwest, covering slightly more than ten miles per day, ascending into the frigid realm above six thousand feet. Unlike the parties of Armitage and Scott, who had struck across the plateau to find the other side, the northern party had a target, a fickle spot that fluctuated on an hourly basis, and had been forty miles to the southeast when Bernacchi had determined the location of the Magnetic South Pole seven years before. As they progressed across the ice sheet, gnawing hunger caused by minimal rations begat fantasies of marvelous banquets that the men would serve to each other when they returned. At 3:30 P.M., on January 16, 1909, the three intrepid explorers reached the location where they determined the Magnetic South Pole to be. They set up a camera with a string to work the shutter, planted a flag, bared their heads, and recorded the photo.

Although they felt "intense satisfaction and relief," David writes, "at the same time we were too utterly weary to be capable of any great amount of exaltation. I am sure the feeling that was uppermost in all of us was one of devout and heartfelt thankfulness to the kind Providence which had so far guided our footsteps in safety to that goal. With a fervent 'Thank God' we all did a right-about turn, and as quick a march as tired limbs would allow back in the direction of our little green tent in the wilderness of snow."

To be back at the coast on February 1, when *Nimrod* had been instructed to begin its search, the men needed to cover sixteen miles per day. Descending over the plateau, they managed to keep the pace and even to exceed it on days when the sail could be used. When they reached Larsen Glacier, however, they fell into several crevasses, including one where both David and Mackay hung at the same time, though not deeply. The final leg over lower Larsen Glacier and the ice foot beyond proved to be the most difficult of the return. Thwarted by deep ice canyons and pressure ridges, desperate for food, ultimately lost, they finally spotted the depot on February 2, but as the party made its way toward it, they were blocked by an ice canyon thirty to forty feet deep. Having been on the march for nearly twenty-four hours, and with only two days' biscuit and a bit of cheese in their store, the famished trio encountered a pair of emperor penguins. The nourishment the birds provided permitted the men to set up camp. The following day was also exhausting, but with fresh meat now at hand they allowed themselves to camp about a mile short of the depot. The following day, February 4, while the men were discussing whether to wait at the depot or to try to pull overland to Cape Royds, *Nimrod* spotted the depot flag and fired a cannon shot as a signal.

The three men bolted out of the tent, turning over their cooker as they went, and gleefully ran toward the sound. Suddenly, Mawson disappeared from sight, having fallen into a crevasse with no harness rope to hold him. In shock, David and Mackay peered into the slot to see Mawson on his back on a shelf about twenty feet down, and only a few feet above seawater that flooded the depths of the crevasse. They lowered a harness but were too weak to pull him out, so the first words that Mackay shouted to the ship when he reached her were, "Mawson has fallen down a crevasse, and we got to the Magnetic Pole." A rescue party was immediately dispatched and Mawson was pulled from the depths, thus ending happily one of Antarctica's most grueling journeys. The men had been on the trail for 122 days, covering 1,260 miles, 740 of which had been times three due to relaying. The usual congratulations, ritual washing, and gluttonous food binge awaited the explorers on board the *Nimrod*.

When *Nimrod* arrived at Cape Royds on February 11, all the parties that had been in the field were back except for Shackleton's southern party and a depot party to Minna Bluff led by Joyce. Joyce's party arrived back on February 20, but the plight of the southern party remained in doubt. Shackleton had left instructions that if his party had not returned by February 25, a rescue party should be readied and sent out on March 1. Great apprehension attended the members of the expedition as the deadline came and went and the southern party still had not arrived.

After the traverse to the Magnetic South Pole, the map of Victoria Land was largely complete. Mawson's surveying had tied down most of the prominent peaks visible from the coast between New Harbour and Terra Nova Bay, although the interior reaches of the mountains remained uncharted. During Scott's *Terra Nova* Expedition (1910–1913), three ground parties would fill in some local details, but the remaining white on the map, in

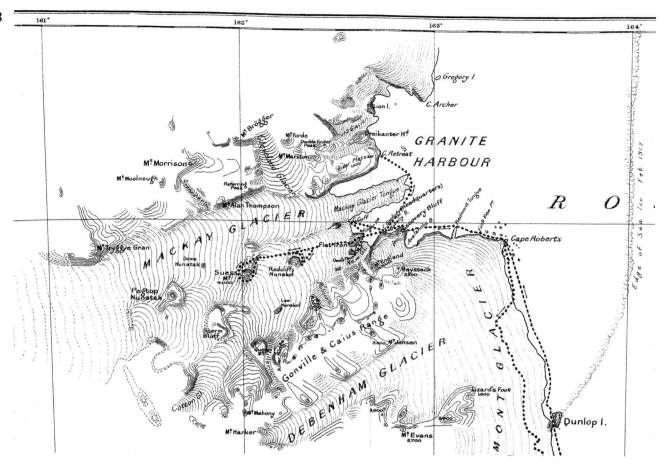

Figure 3.13. Taylor's map of the Granite Harbour area plots the route taken by the party as it man-hauled along the piedmont glacier and climbed Mackay Glacier to Mount Suess.

particular the interior of northern Victoria Land and the two major dry valleys (Wright and Victoria) to the north of the Taylor, would not be mapped before the advent of aerial photography and the wave of ground parties that spread throughout the Transantarctic Mountains following the International Geophysical Year (1957–1958).

The first of the *Terra Nova* parties included Griffith Taylor (geologist and leader), Frank Debenham (geologist), Charles Wright (physicist), and Edgar Evans (petty officer and a member of Scott's 1902–1904 expedition). In January 1911 this party man-hauled up Ferrar Glacier and down into Taylor Valley, reaching a point on the eastern side of Nussbaum Riegel, where they could see the ocean. From their farthest point near the mouth of the valley, they were able to survey local landmarks and tie them back to Ross Island. From there they retraced their route back to the head of Taylor Valley, where they determined correctly that, of the four glacier arms, the western arm that Armitage and Scott had taken to the interior was the head of Taylor Glacier, and the southern arm was the head of the Ferrar Glacier, with the two glaciers shouldering each other in the broad névé to the south of Knobhead before turning 90° and flowing down their respective valleys (see Fig. 2.10). On their way to Blue Glacier via the Ferrar, they chose to bypass Descent Glacier, an experience that Evans, a member of the Armitage A team, was not eager to repeat. From the mouth of Ferrar Glacier, the party explored the upper reaches of Blue

Glacier and the western side of Koettlitz Glacier, mapping details of the Royal Society Range and the mini–dry valleys that make up the foothills to the east of Blue Glacier (see Fig. 1.17).

The following year Taylor led a geological party, including Debenham, Tryggve Gran, a Norwegian with ski experience, and Robert Forde, one of the petty officers, to Granite Harbour (Figs. 3.13, 3.14; see Fig. 1.14). From a camp about ten miles up Mackay Glacier, they climbed the three thousand–foot summit of Mount Suess, sighted and named by David's party three years earlier, with a view to the inland plateau (Fig. 3.15). On the moraine downstream from Mount Suess the party discovered fossil fish plates in loose blocks of shale, which upon return to Australia would be identified as Devonian in age. The party mapped the area geologically, although the men did not find any of the fossil-bearing shales in outcrop. They also added several colorful names to the Antarctic lexicon. Gondola Ridge, forming the spine of the nunatak with Mount Suess at the summit, evoked the outline of a Venetian gondolier. Sperm Bluff and Killer Ridge were thought to resemble the blunt nose of a sperm whale and the profile of a killer whale, respectively (Fig. 3.16). And finally, Queer Mountain was peculiar because it was completely surrounded by glaciers and displayed every one of the area's rock types in its outcrop.

The third of the *Terra Nova* parties included Lieutenant Victor Campbell (leader), Raymond Priestley (geologist), Dr. Murray Levick (surgeon), George Abbott and Frank

Figure 3.14. This view into Granite Harbour toward the southeast centers on the heavily crevassed Mackay Glacier where it lets loose of its bed and flows forward as the Mackay Ice Tongue. Discovery Bluff is the dark ridge under cloud in the middle ground. Taylor's party avoided the main icefalls of Mackay Glacier by following The Flatiron, the thin ridge in front of Discovery Bluff, to an elevation above the heavily crevassed stretch. Fig. 1.15 was shot from the left (distal) end of the ridge in the direction of this image.

Figure 3.15. This panorama, looking to the southwest, encompasses the area surveyed by Taylor's party in 1911–1912 (compare Fig. 1.14). Following counter-clockwise from the ridge on the right margin, the features include: (1) the eastern end of the ridge containing Referring Peak; (2) Gondola Ridge and Mount Suess almost entirely obscured in cloud; (3) Sperm Bluff, with its left end reminiscent of a sperm whale rising, a steep profile on the left and more gentle incline to the right; (4) Queer Mountain, the buttress to the left and behind Sperm Bluff, partially mantled in cloud; (5) the southeastern end of Killer Ridge, with its snow-clad ridgeline that extends to the left rear (Fig. 3.16 looks back to this ridge and Sperm Bluff from the southwest); (6) Redcliff Nunatak, the dark island free of snow; (7) a name-less buttress on the north side of the Gonville and Caius Range; and finally (8) the ruffled pattern of cre-vasses at the head of the falls into Granite Harbour (approximately the right edge of Fig. 3.14).

Figure 3.16. The southwest wall of Killer Ridge rises eight hundred feet up from Miller Glacier. The lower heights are composed of granite, the upper reaches dolerite, intruded along an old erosion surface on the granite. A shoulder of Queer Mountain holds the left foreground.

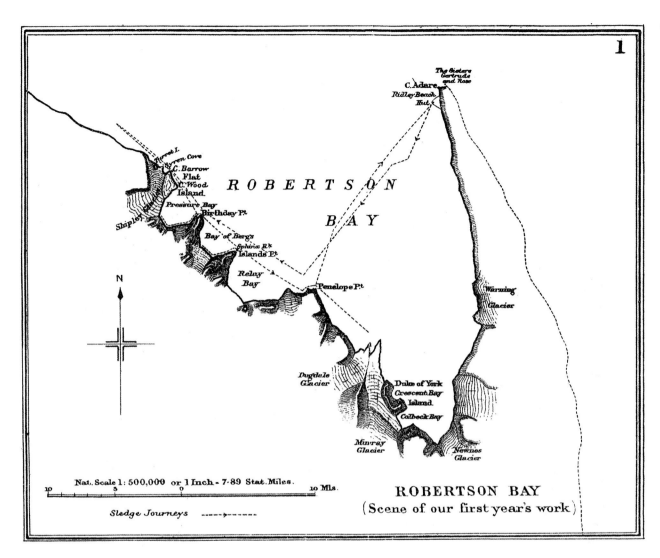

Figure 3.17. After landing at Ridley Beach on the tip of the Adare Peninsula in February 1911, Campbell's party had no more success than Borchgrevink in escaping the steep rock walls and cascading
ice falls around the perimeter of Robertson Bay and down the seaward wall of Cape Adare. After
wintering over, the party was picked up and moved by the *Terra Nova* to Terra Nova Bay.

Browning (petty officers), and Harry Dickson (able-bodied seaman). These men survived
one of the most horrendous of Antarctica's sagas. The *Terra Nova* had landed them at
Cape Adare on February 18, 1911, at the site of Borchgrevink's hut, for a winter-over. They
soon found that the steepness of the mountains and the instability of the sea ice prevented
them from exploring beyond the confines of Robertson Bay, exactly as Borchgrevink had
reported (Fig. 3.17; see Fig. 1.6).

On January 3, 1912, *Terra Nova* picked up the party and sailed south to Terra Nova
Bay to drop them for a month of mapping. The ship was able to moor against fast ice in
the bay between the Northern Foothills and the Southern Foothills—later named Inex-

Figure 3.18. In this high-elevation image of Terra Nova Bay looking southwest, seasonal ice has begun to break up. The light-colored bank of ice on the right side of the image is sea ice that has persisted for a number of seasons. Embedded within this is the Campbell Ice Tongue issuing in from the right. The snow-covered ridge in the middle ground is the Northern Foothills. Immediately to the left (south) of this is the dark face of Inexpressible Island, where Campbell's party wintered over in 1912. Hell's Gate is the name given to the gulf between these two features. Reeves Glacier with its two prominent nunataks is in the center rear of the picture. In the upper left corner Backstairs Passage cuts the dark, narrow outcrop of rock. Mount Crummer is the small peak to the left (south) of Backstairs Passage.

pressible Island (Fig. 3.18; see Fig. 3.12). For the remainder of January and half of February, the men explored around the northern end of Terra Nova Bay, mapping along the base of the towering mountains sighted by David's party (Eisenhower and Deep Freeze Ranges of today), but they were able to penetrate only about ten miles up Priestley Glacier (Fig. 3.19; see Fig. 3.11), so were unaware that this is the most northerly of the outlet glaciers that drains across the mountains from the polar plateau.

When the *Terra Nova* failed to return that season, the party faced a winter-over with minimal rations and tents that were all but shredded by the wind. The men survived by digging a snow cave in a hard drift, lining it with dried seaweed for insulation, and killing a number of seals for a meager food supply during the winter. Conditions were squalid and miserable in the ice cave on Inexpressible Island, but the men's temperaments were buoyant. When the frail survivors emerged from their cave in the spring, hope had prevailed, and they were determined to rescue themselves by man-hauling back to Scott's hut at Cape Evans, following the coastal route blazed by David, Mawson, and Mackay.

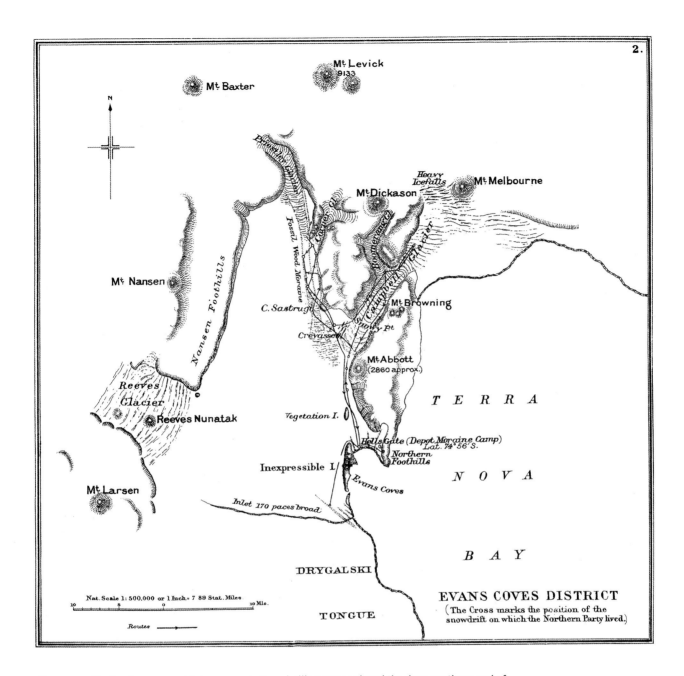

Figure 3.19. During January and February 1912, Campbell's party explored the deep northern end of Terra Nova Bay. They correctly mapped the mouth of Priestley Glacier but misunderstood the head-reach of Campbell Glacier. In fact, this glacier does not arise on the side of Mount Melbourne but descends from a subtle divide, with most of the flow coming from Boomerang Glacier (see Fig. 3.11). Campbell Glacier, as it is mapped today, drains the south side of Mount Melbourne and flows as a graceful ice tongue into the northern end of Terra Nova Bay (see Fig. 3.18).

Because it was early in the season with everything frozen tight, Campbell's party encountered neither the slushiness nor the wet blizzards that had been such a drag on David's progress.

On October 26, camped at Depot Island, the party picked up the letters left by David's party, as well as all of the rock specimens, and continued down the coast. At Cape Roberts at the southern entrance to Granite Harbour, the men found the depot left by Taylor's geological party six months earlier, and after months, their hunger was temporarily sated. Priestley writes, "The day and night of the 29th merged into one glorious feast, and when we started again on the following morning our mouths were sore from nibbling biscuit, and pretty well three day's sledging allowance had been accounted for. In fact, I had served out a week's butter, raisins, and lard amongst the six of us, and the only thing we carried away with us externally was a small piece of butter and lard each." On November 1 the party picked up another cache from Taylor's party at Cape Bernacchi, which included pemmican in addition to more raisins. From Butter Point they took a southerly route around McMurdo Sound to stay on solid ice.

Their arrival at Hut Point was greeted with crushing news when they read in a note left in the hut that Scott and all his party had perished the previous year, and that a search party had set off to the south only a week before. Campbell's party had not pushed far into the white spaces of the map, but their gaze into darkness was perhaps deeper than any party before or after. They had stood on the banks of the Styx and for many months had seen clearly the other side. They had stared down Death through their determined goodwill and camaraderie, only to be cut to the heart by this news of their lost comrades. Such was their conflicted spirit as they trudged the final leg back to the base at Cape Evans for the survivors' ritual washing and food.

4 Penetrating the Interior
Discoveries in the Central Transantarctic Mountains

On February 25, 1909, the day that Shackleton had given instructions to begin preparations for a search party, he and his three sledge mates, Wild, Adams, and Marshall, were camped fifteen miles north of Minna Bluff. The men had barely survived a seventeen hundred–mile march on starvation rations when, two days earlier, they had tenuously reached the Bluff and its well-stocked depot (Fig. 4.1). In addition to an abundance of the usual sledging rations, biscuit, pemmican, cocoa, and tea, the cache contained eggs and freshly cooked mutton from the *Nimrod,* as well as gifts, including "Carlsbad plums, cakes, plum puddings, gingerbread, and crystallized fruit." The men sated themselves, crawled into their sleeping bags with little stashes of goodies, and awoke to a breakfast "consisting of eggs, dried milk, porridge and pemmican, with plenty of biscuits." They marched through the day, stopping once for lunch, and camped at 8:00 P.M.

When Shackleton rousted the second tent on the morning of the 25th, he found Marshall cramped so badly from dysentery that he could not stand. Although food was no longer an issue for their survival, there was urgency in reaching *Nimrod,* since once a search party was stationed at Hut Point, the ship had been given leave to depart for New Zealand provided McMurdo Sound had begun to freeze over. Before any decision could be made, a blizzard came down and pinned the party in their tents.

At 1:00 A.M. on the 26th the storm broke. Marshall was able to walk, but not to pull, as the men headed out three hours later. Fortified by stops for lunch, tea, and two hooshes at 7:00 and 11:00 P.M., the party covered twenty-five miles before turning in at 1:00 A.M. on February 27. Three hours later they were up again and marched until 4:00 P.M. Mar-

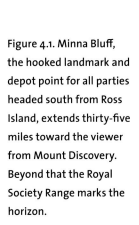

Figure 4.1. Minna Bluff, the hooked landmark and depot point for all parties headed south from Ross Island, extends thirty-five miles toward the viewer from Mount Discovery. Beyond that the Royal Society Range marks the horizon.

shall's dysentery had worsened to the point that Shackleton decided to leave him with Adams and to push on with Wild for the ship. Lightened to only a compass, their sleeping bags, and one day's ration of food, Shackleton and Wild sledged until 9:00 P.M., when they had a hoosh, then continued their march until 2:00 A.M. They stopped for an hour and a half but were unable to sleep, so they continued trudging into the morning. By 11:00 A.M. the food was gone. At 2:30 P.M. they spotted open water four miles south of Cape Armitage. Then a blizzard came down. While groping along with minimal visibility, they suddenly came up on the ice edge flexing rhythmically from the roll of the waves.

With the expectation of finding the search party at the *Discovery* hut, and fearful they would not be able to reach land at Pram Point (the south side of Observation Hill) due to the low visibility, Shackleton decided to leave the sledge and sleeping bags and to make for Castle Rock, the landmark from which there would be a fairly straightforward descent along the peninsula to Hut Point. When they reached the ridgeline, they looked down into open water and no sign of the *Nimrod* (see Fig. 2.7), nor as they approached the *Discovery* hut was there any sign of life as they had expected. A note said that the ship would have waited by the Erebus Ice Tongue until February 26, but it was already the 28th. The exhausted men found ample food, but without their sleeping bags they spent a miserable night sitting wrapped together in a swath of roofing felt. In the morning as the sunlight warmed them, they managed to tie a Union Jack to Vince's Cross at the end of Hut Point and to set fire to the magnetic hut in hopes that these would attract the attention of the *Nimrod*.

At that moment *Nimrod* was sailing south to drop the search party, and Wild spotted her in a mirage. The men flashed the ship with the heliograph, and by 11:00 A.M. they were on board, with great rejoicing by all hands. Immediately Shackleton ordered a rescue party to be organized, and then ate his fill. At 2:30 P.M. he was out in front of a four-man party including Mawson, Mackay, and McGillon (who had come down that season on the *Nimrod*); the men marched till 10:00 P.M., ate, cat-napped, and were on the trail again at 2:00 A.M. At 1:00 P.M. on March 2, they reached Adams and Marshall. Marshall had recovered enough that he could even pull a little. They packed up and headed back, camping at 8:00 P.M. Up at 4:00 A.M., they reached the ice edge at 3:00 P.M., with no ship in sight (again), and ice visibly forming in the sound. Abandoning all but their sleeping bags and the geological specimens, the party made its way along the ice edge to Pram Point, where they scuttled the rocks, and taking only sleeping bags, climbed around Crater Hill and down to Hut Point by 9:50 P.M.

After the others had eaten and turned in, Shackleton and Mackay lit a carbide flare on the point, where it was spotted by the *Nimrod* standing off about nine miles. By 1:00 A.M. on March 4, everyone was safely on board. Lest anyone ever again doubt Ernest Shackleton's physical stamina, he had slept no more than fifteen hours in the previous six days and had trekked and hauled more than ninety miles, all coming off a seventeen hundred–mile odyssey of discovery on the brink of starvation.

The odyssey had actually begun seven years before, with great jubilation, as the southern party (Scott, Wilson, and Shackleton) drove dog teams away from the *Discovery* on November 2, 1902, bound for the south to the limit of their strength and resources. Earlier that spring Scott, Shackleton, and Feather had driven dogs to the end of Minna Bluff, where they made a depot consisting of six weeks' provisions for three men and 150 pounds of dog food. The hooked end of Minna Bluff (see Fig. 4.1) straightened into a perfectly linear ridge that lay to the north-northwest, directly to the center of Mount Discovery. They placed the depot about eight miles from the end of the Bluff so that the last volcanic cone at the terminus aligned with the summit of Mount Discovery. This connected to the main mass of the western mountains that arched southwest and then south. At the southern termination, the land appeared to be separated into two islands with the horizontal line of the Barrier bridging the gaps. From his vantage at the depot Scott thought that he could see the termination of the mountains to the south, with the Barrier surface extending around them toward the northwest. At the time, no one knew whether the western mountains constituted an archipelago of islands or were part of a continental landmass; Scott favored the island hypothesis.

Because of this impression, the southern party anticipated a long, boring haul over the Barrier surface, but that direction was into the deep unknown, with the ultimate prize the South Pole, though if even the most favorable scenario played out, it was unlikely that they could reach the pole itself. Armitage, in plotting a route to the Magnetic South Pole, chose to attempt a crossing through the mountains rather than trying an end run to the south. It thus fell to a party led by Barne to explore the mountains to the southwest of the Bluff after the southern support party had returned to *Discovery*.

The twelve men of the support party, man-hauling five sledges as a train, had left the ship two days before the southern party, which caught up with them the first evening about fourteen miles south of White Island. Because the dogs were performing well, weight was shifted to their sleds, and the parties continued on more and less together, with the southern party arriving at the Bluff depot on November 10 and the support party a day later. On the morning of November 12, the groups set out together, shifting more weight to the dog teams to equalize their speeds. By evening the merry explorers were camped farther south than Borchgrevink's previous record. In the morning a group photo with all flags flying commemorated the moment, before six of the support party turned back. After two more days of pulling, the southern party filled up their loads to more than two thousand pounds distributed among six sleds, and the remaining six men of the support party turned back.

Scott, Wilson, and Shackleton were now poised to advance to high southern latitudes, where answers to the questions of the continental nature of Antarctica might be found. But almost as soon as the support party left, the pace of the southern party began to slow. The dogs did not seem to have the same vigor, and the air temperature was balmy (for Antarctica), causing the snow to become sticky. Within a few days it became necessary to ferry the loads, and the daily distance averaged five or six miles even though the march was fifteen to eighteen. During this period a thin mist hung in the air much of the time, producing fine ice crystals that settled out of the sky, adding to the soft snow surface. Although the drift of ice crystals impeded progress, it created the most extraordinary atmospheric effects concentrating light in compound arcs, halos, and parhelia around the sun, sometimes with faint colors of the rainbow. On the morning of the 19th, when the air cleared a bit, the men saw mountains off to the southwest, farther south than they had seen before. It gave the impression of several detached portions of land with broad gaps separating them, and possibly further islands at the southern end of Victoria Land.

On November 20 the party advanced only three and a third miles, on the following day, four. With clear skies more mountains appeared farther in the distance to the southwest. By this time it was painfully clear that they would not be able to make any great penetration into the interior. With no certain end in sight to the mountains, the party decided to strike toward land in order to survey these new features and possibly to collect geological specimens. This cheered Wilson, because he was eager to record the mountains in his sketchbook. The intention was to establish a depot close to some landmark, and from there to explore south along the mountain front with much lighter loads.

On November 25 the party crossed 80° S latitude. Scott noted, "All our charts of the Antarctic Regions show a plain white circle beyond the eightieth parallel; the most imaginative cartographer has not dared to cross this limit, and even the meridional lines end at the circle. It has always been our ambition to get inside that white space, and now we are there the space can no longer be a blank; this compensates for a lot of trouble."

For the next three weeks the party inched toward the mountains at an agonizingly slow pace, gradually moving onto a night schedule, relaying all the while. During that time it was the dogs, not the men, that called a halt to the marches. The teams seemed simply not to have the strength. Upon reflection the men concluded that the problem lay in the dog food, a meal of dried fish, which unbeknownst to anyone had apparently spoiled during the passage through the tropics. Isolated, the men could do nothing but push on with the sick animals, feeding them dog meat when one of their fellows dropped. The men never ate dog meat, and it probably would have been better for them if they had; traveling on extremely minimal rations, they talked and dreamed of food incessantly.

Wilson wrote in his diary of this time, "Dreams as a rule of splendid food, ball suppers, sirloins of beef, caldrons full of steaming vegetables. But one spends all one's time shouting at waiters who won't bring one a plate of anything, or else one finds the beef is only ashes when one gets to it, or a pot of honey has been poured out on a sawdusty floor."

The only joy during these grueling days was the increasing detail that came to light in the mountains. A major range with an east-west strike took shape to the west. The highest summits, which Scott estimated to be more than ten thousand feet, descended to the east through a system of ridges to a snow-covered plateau sprinkled with low peaks. To the front of these was a low, rolling snow slope that merged with the Barrier. On its southern flank, the Britannia Range, as it was called, rose in high, bare-rock cliffs on the northern side of what appeared to be a major strait through the mountains. This breach was fifteen or twenty miles wide, but its far end was not visible beyond the mountains on the south side of the passage (Fig. 4.2).

These mountains to the south appeared to be high, to run north-south, and to be snow covered, but their distance obscured details. In contrast, the cape on the south side of the strait rose clearly as a broad snow dome and appeared to be connected in the distance to the mountains. On December 14 a distant peak and a rocky patch on the promontory (Cape Selborne) were chosen to be the alignment for placing Depot B. As soon as the men had secured the camp, they headed toward the rocky patch to see what sort of rock it was.

Having hiked less than a mile in cloudy, low-definition conditions, the men were stopped by a huge chasm in the ice. Wilson recorded, "It was a wonderful sight, a chaos of ice masses jumbled up in crevasses of 40, 50, 60 feet deep, the valleys some hundreds of feet across full of tumbled blocks and frozen pools of water." Clearly there was no way to cross this divide, so the men returned to camp. The following day with good light, Shackleton took some photos of the chasm, but the men found no routes across, so with a lightened load the party started its trek south along the mountains.

Two sledges carried four and a half weeks of food, with three weeks' worth left at Depot B to see the party back to the depot at the Bluff. Without the need to relay, morale was high, but the dogs were spent, and the previous month had taken a severe toll on the physiques of the men. With no fixed goal other than to reach south to the limit of their endurance, the party pushed forward (Fig. 4.3). Each day brought new sights of lofty peaks beyond rolling, snowy foothills (Fig. 4.4). In general, the foothills were several thousand feet high, so they obscured their connections to the more distant mountains that peeked out from behind. New summits that appeared faintly on the horizon during clear spells lured them on. Scott wrote, "Ever before us was the line which we were now drawing on the white space of the Antarctic chart. Day by day, too, though somewhat slowly, there passed on that magnificent panorama of the western land. Rarely the march passed without the disclosure of some new feature, something on which the eye of man had never yet rested; we should have been poor souls indeed had we not been elated at the privilege of being the first to gaze on these splendid scenes."

By December 22 the party had reached a portion of the coast where a breach in the foothills allowed a view up into the face of the main range. Beaumont Bay, as the opening was named, was fed by a glacier that arose high on the side of the mountain, which Scott

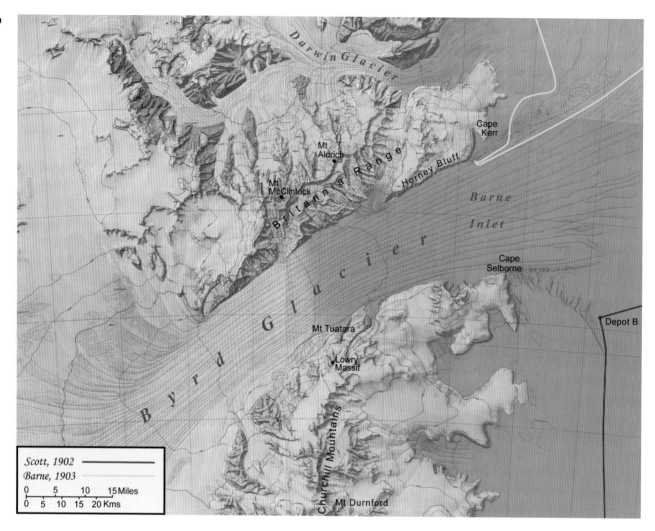

Figure 4.2. Shaded-relief map of the Byrd Glacier area. In early December 1902, as Scott's party hauled toward the mountains, the ice shelf retreated back into an inlet that appeared to cut clean through. On the north side of this inlet, sheer walls of the Britannia Range climbed to heights greater than ten thousand feet. The party approached Cape Selborne, where it placed a depot for the return. As the men hiked toward the rocky outcrop at the cape, they were stopped by a great chasm in the ice shelf. Thereafter they traversed south along the mountain front. The following year as Barne's party tried to approach rock on the north side of the inlet, it too was thwarted by an immense chasm in the ice. It is now known that Byrd Glacier carries by far the largest volume of ice of any of the outlet glaciers that cross the Transantarctic Mountains, and that these chasms are where it rips into the Ross Ice Shelf on its way to the sea.

speculated (correctly) "form(s) the backbone of the whole continent." The upper slopes reaching to more than nine thousand feet were blocky, with snow-clad, horizontal layers, similar to the Royal Society Range. A series of steep spurs dropped from the summits that stood above the high ridgeline. The peak directly opposite the bay formed a perfect pyramid, and so was named Pyramid Peak (Fig. 4.5). The highest portion of the range

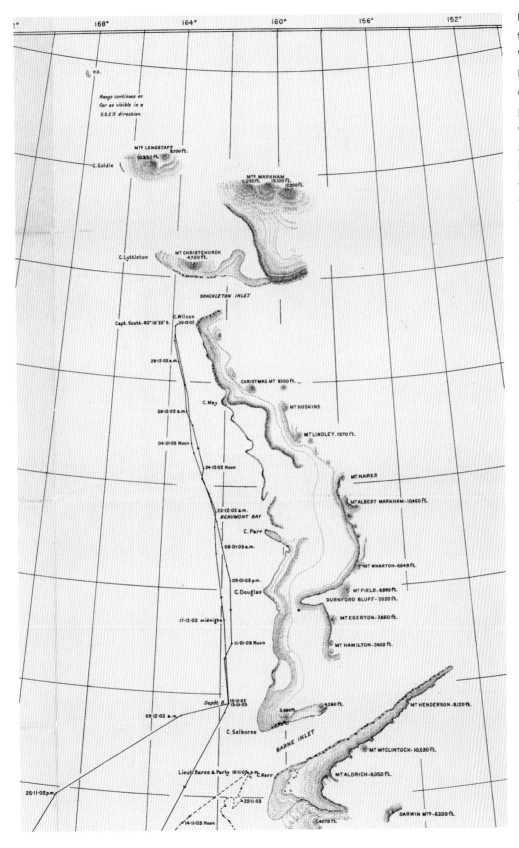

Figure 4.3. Scott's map of the sledging journeys from Winter Quarters at Hut Point. Note the change of course by the southern party in late November, when the men decided to abandon their southerly heading and to attempt to reach rock to the west. When the party met the ice chasm to the east of Depot B, it continued south along the mountains until reaching Shackleton Inlet, which Shackleton recognized six years later to be the mouth of Nimrod Glacier.

Figure 4.4. (above) The central Churchill Mountains, viewed in this scene from Mount Durnford, were first seen by Scott's southern party in mid-December of 1902. Mount Durnford also served as one of the best-situated stations for the Topo South survey party in 1961 (Chapter 7).

was a flat-topped massif with a horizontal layer of dark rock exposed at the summit. This feature was named Mount Albert Markham, for the brother of Clements Markham, and the man who had first introduced Scott to Sir Clements. Some miles to the south a conical peak, standing prominently out from the main range, served as a landmark for the next several days.

Up until this time, although the men were ravenous and losing weight, they continued to be fit for pulling; however, at this camp in his regular examination of the men's gums and legs, Wilson noted a deterioration of Shackleton's gums, the first sign of possible scurvy. Wilson confided this worrisome news to Scott, but the two kept it from Shackleton for some days to come.

For weeks the men had planned to celebrate Christmas Day with a full ration of food, and as the day approached they "discussed and rediscussed" just what that would be. Breakfast, cooked for an extra fifteen minutes on the primus so that it was steaming hot, included a full pot of "biscuit and seal liver, fried in bacon and pemmican fat," followed by a large spoonful of jam. Their spirits lifted by the added calories, the men received the Christmas gift of a harder snow surface as they started their morning march under a cloudless, windless sky. At lunch they savored a cup "of hot cocoa and plasmon with a whole biscuit and another spoonful of jam."

When the men pitched camp for the evening, they had logged eleven miles, the longest daily distance traveled in many weeks. The hoosh that night had "a double 'whack' of everything" and was so thick that a spoon could stand up in it. Boiling cocoa followed, and then Shackleton produced a lump of plum pudding about the size of a cricket ball and an artificial sprig of holly that he had ferreted in the toe of a sock until this moment. The incessant conversations about food gave way that night to thoughts of family and friends and what the English Christmas might be like that year.

South of Beaumont Bay the rolling foothills had again risen from the ice shelf to elevations of several thousand feet, obscuring the high mountains farther west. At places the ice fell over the steep edges of the foothills in frozen waterfalls up to one thousand feet high. Elsewhere rocky promontories jutted out from the snowy rise. As the party progressed south, a high peak began to materialize near the coast at a great distance. Calculations suggested that it was beyond 83° S. This southernmost sighting was named Mount Longstaff after the man whose early, generous contribution of £25,000 had floated the expedition (see Fig. 4.3).

Even though he wore sun goggles, for a number of days Wilson had suffered excruciating sun blindness, owing to his faithful sketching of the mountains with unprotected eyes. On December 27 the condition was so bad that he covered his eyes completely and marched blind. This was a day of exciting discovery, which Scott described to Wilson as they moved along. The snowy hills to the west terminated in a prominent, snow-capped, rocky cape. Beyond the cape the ice surface withdrew into a broad reentrant, which appeared first as a bay and then, as the day progressed, a strait that cut deep into the mountains, similar to the feature to the south of the Britannia Range. New mountains appeared on the south side of the strait as the party continued to work abreast of the cape. Starting as low foothills to the east, they rose higher and higher to the apex of a massif far grander than any seen on the entire journey. With an intricate system of ridge-

Figure 4.5. (opposite bottom) On December 22, as Scott's party drew abreast of Beaumont Bay, the men were able to see past the foothills and up into the main range. Pyramid Peak stood at the head of the embayment. In this view taken from south of Beaumont Bay, looking to the northwest, Pyramid Peak centers the image while the black-topped massif to the left (south) is Mount Albert Markham.

Figure 4.6. A symphony of ridgelines crescendos up the northern face to the summit of Mount Markham. Nimrod Glacier flows from right to left around Kon-Tiki Nunatak, the dark island in midstream. In this view taken from the northwest about forty miles up glacier, the relief from the edge of the glacier to the summit of the massif is more than eleven thousand feet. Scott's view would have been from the northeast at the mouth of the glacier, but the ridge systems would have appeared similarly.

lines rising to over thirteen thousand feet, it pitched most steeply just at the top where it leveled onto a flat surface with two prominent peaks on its crown. Scott described it as "a giant among pigmies" and named it Mount Markham, after Clements Markham, the father of the expedition (Fig. 4.6).

On December 28 the party pulled another six miles south to beyond 82° S, stopping in the afternoon to survey and sketch the new mountains. From this camp, which was their farthest south, they planned to continue on skis in order to gain a view up the strait to see whether the ice continued as a level surface to the horizon or rose in a glacier to the interior. Was Mount Markham part of an island? Was it separated from the mountains that they had been surveying for the past month by a glacier? Were the mountains continental or a volcanic archipelago?

Once again Antarctica held her secrets. On December 29 the party was pinned in its tent by a blizzard. On the 30th a fog settled in. Scott and Wilson skied out from camp for a mile or so but turned back for fear of losing their way. On the last day of 1902 the cloud pulled back, allowing them to ski toward the south, but as they cleared Cape Wilson, the western reaches of the strait remained shrouded in fog. The snow surface pulled north around Cape Wilson, with no land visible farther west on the north side of the strait. On the south side of the strait, the Mount Markham massif ended in a steep bluff (today known as Inaccessible Cliffs). Without the view into the strait, the last hope of answering the question of the character of the mountains lay in reaching bedrock at Cape Wilson. However, as the party approached the cape, they encountered another deep rift in the ice, similar to the one that had turned them back at Cape Selborne, so they failed to collect a geological specimen that would have provided an answer.

The time had come to turn and head north. They had set the records: the longest assault on either pole, the southernmost advance on Earth, and 250 miles of new mountains charted, with every expectation that they continued beyond Mount Longstaff into the heart of the Antarctic. With two weeks' rations and a desperate hope for fair weather, the men would literally be pulling for their lives.

ASCENT OF MOUNT MARKHAM

In Antarctica, first ascents generally fall into two categories: the easy way, set out on the summit by a helicopter, and the old-fashioned way, one step after another till you reach the top. I've done plenty of both. It comes with the territory.

However, my most gratifying first ascent, that of Mount Markham, falls somewhere in between. During the 1985–1986 season, I was working out of a large, helicopter-supported field camp at the head of Lennox-King Glacier. Mount Markham is a huge buttress with intricate ridge systems rising up its north, east, and west sides (Fig. S.7). To reach the summit from any of these ridges remains a challenge for extreme mountaineers. The south approach, on the other hand, looked easy to me, at least by comparison. The summit morphology of the Mount Markham massif is that of a three-sided escarpment or buttress surrounding a flat area about a half-mile wide. The flat, which sits at about twelve thousand feet, descends gently to the south over a distance of thirteen miles to a saddle at about nine thousand feet, one thousand feet lower than the normal, allowable ceiling for flying helicopters in the field.

The others in the party were Russell Korsch, a geologist from Australia, and David Edgerton, my graduate student. Our plan was to be put in by helicopter in the saddle along with our snowmobile and sled, and then to drive to the top. The day of the put-in was still and clear. Once the helicopter had departed, we quickly packed the sled and started up the long ramp. Russell and David rode on the loaded Nansen sled, and I drove. With the combination of the weight, the incline, and fairly soft snow (in which we left a two-inch deep track), the snowmobile crept for most of the distance at full throttle.

Figure S.7. Aerial view of Mount Markham from the east. The path followed by the snowmobile is shown in pink. The summit of Mount Markham is the dark, pyramidal peak to the right of the end of the snowmobile path.

Because air thins with increased elevation, we had to rejet the carburetor every thousand feet to maintain the proper air-to-gas mixture. The jet was a thin metal tube about three quarters of an inch long that screwed into the throat of the carburetor. I loosened and tightened it with a special little wrench about two inches long, and when I took it out or set it in place, I needed a bare finger and thumb to hold it, gloves being too bulky to fit into the throat of the carburetor. The whole procedure took about eight minutes, and generally I had to put my bare hand back into a mitten once or twice to regain feeling.

Although the day had started out calmly, after several hours a headwind sprang up, flowing directly down on us out of the north. To make matters worse, a ceiling of cloud began to form as the wind quickened, settling in at about eleven thousand feet. I stopped about one hundred feet below the cloud to do the last rejet. Although we could still see rock outcrops that bounded the ramp

Figures S.8, S.9. Panorama to the north from the summit of Mount Markham.

we were climbing, all surface definition on the snow at our feet was gone. I worried that if the cloud was too thick, I would have no visual means of staying in the middle of the ramp.

As I started the rejetting, an eerie incident occurred. Out of nowhere, a skua appeared, circled us several times, landed on the snow about fifty feet up slope, and sat there calmly watching me. The powerful bird, more than five hundred miles from open water and its source of food, squatted inscrutably beneath the gathering gloom. Then, having rested, it spread its tawny wings, rose slowly into the wind, drifted east beyond the rocky edge, and dropped from sight.

If ever there was an omen, this surely had to be. But was it bad or was it good? I took it to be good, and fired up the snowmobile for the final leg, about a mile to the summit flat. When we hit the bottom of the cloud, all visibility and surface definition went to zero. I steered by keeping the wind directly on the nose

Ascent of Mount Markham

of the snowmobile, and figured that when the slope went flat, we would be on top. To make matters worse the wind speed steadily increased as we continued to climb. Then a blessing, the snow surface began to break up into patches of blue ice, and I again had a surface that I could see, albeit for only about thirty feet ahead.

When the surface finally flattened, the patches were less snow than blue ice, and the wind was an unwavering stream at probably thirty to forty knots. I chose a patch of snow to plant the tent. Russell and David were stiff from the cold; I was sweating from gripping the handlebars of the snowmobile behind its windscreen. The next ten minutes were a frantic blur, setting up the Scott tent "by the book" in a strong wind. With shelter secure, we dove into the tent, started the stove, and melted some snow for a brew. Dinner followed. Although the wind was still howling as we crawled into our sleeping bags, the only real concern as we fell asleep was, when would it end?

The answer came as we awoke to a brilliant sun and not a breath of wind. Our marching orders being "Geology First," we carefully worked down the ridgeline that drops from the northwest corner of the buttress and spent the day mapping on the exposed cliffs. After coming back to the base camp, we drove over to the summit rise, parked the snowmobile, and climbed a couple of hundred feet up a slope of snow and sandstone ledges. The summit of Mount Markham is a broad, flat slab of sandstone maybe one hundred feet across. With only the slightest bit of breeze, we walked to the northern edge of the slab and gazed out at the most spectacular panorama I have ever beheld (Figs. S.8, S.9). A grand and graceful system of ridgelines rose twenty-five miles out and steadily climbed through highs and lows, converging on our lofty spot. Nimrod Glacier cut the scene in half, with mountains on the other side extending back another fifty miles or more. Both the Ross Ice Shelf and the Polar Plateau were in view.

Can it ever be any better than this?

By then completely drained, three men and ten dogs started back on New Year's Day, 1903. On January 3, two dogs dropped in their harnesses, on the 4th another. By January 7 the remaining dogs were beyond pulling at all, so they were cut free to follow along behind. Storms with melting snow pelted the party, but the men dared not stop short of Depot B. A sail rigged to the sledge aided the march when southerly breezes were blowing. During one blizzard the men sailed so fast that they could barely control the sled. As they neared the area of Depot B on January 12, another blizzard came down, totally obscuring the landmarks by which they could reckon their way. They marched blind for a distance and then camped. The next day brought the same gray weather and another blind march, but after three hours they camped again, afraid that they would pass the depot. In the afternoon during a brief clearing, Scott spotted the flags of the depot about two miles away, and they reached it two hours later.

That evening a "fat hoosh" brought brief comfort to the men. On the 14th they shook down their sledges, with three week's rations to cover 130 miles to Depot A. In the previous week, Shackleton had developed a troublesome cough and was beginning to suffer from the scurvy, but on this day a medical examination revealed the first signs of scurvy in Scott and Wilson too, though their performance had not yet been impaired. Wilson was so alarmed by Shackleton's health that Scott forbade him from participating in any man-hauling or camp work, lest he break down. So Shackleton trudged along in his harness without pulling, then went it alone on skis, and sat discontentedly while Scott and Wilson set up and broke down camps.

Slowly the old landmarks began to reappear, Mount Discovery, Minna Bluff, Mount Erebus. For a number of those days the party marched through fog in total whiteout conditions, always with "a desperate hunger," as Shackleton's condition worsened. On January 26 the men were heartened to come upon tracks from Barne's party returning from his reconnaissance of the coastline to the west. Two days later they reached Depot A, knowing that they had survived.

At last the gaunt trio had food aplenty, plus a sack brought out from the ship with letters and special treats for everyone, including a packet of tobacco for Scott, the only smoker. The three filled themselves that night with a triple serving of hoosh topped off by "the thickest brew of cocoa with 'lashings' of jam and biscuit." Because of his state of health Shackleton did not overeat, but Scott and Wilson gorged themselves. In a short time the dried food in their bellies began to expand, causing so much discomfort that they feared they might actually burst, and they were forced to walk in a circle around the tent until the pressure in their stomachs subsided.

On January 29 another storm was blowing, so the party laid up for the day, then pushed on for the final stretch back to the ship. On February 3, six miles from their goal, they were excitedly met by Skelton and Bernacchi, who escorted them back to *Discovery* and a hero's welcome.

Having been unable to see into either of the major straits that they had passed, Scott was anxious to hear the report from Barne of his explorations to the southwest of Minna Bluff. After leaving the southern party, the support party returned to *Discovery,* where they made ready for another traverse to the Bluff depot. On New Year's Day the party of six, led by Barne, was back at the depot, poised to sledge to the southwest with rations for twenty-four days. According to instructions, Barne was to set a course to the southernmost appearance of land—that is, to the south cape of the second island as viewed from the Bluff. On January 3 a third island appeared farther to the south, and as the day's march continued, it extended into an elongate coastline. On the 4th the party advanced in a snowstorm with no visibility following a compass setting toward the new southernmost land. At lunchtime on the 5th, as a clearing blew through, Barne could make out far to the south a new "indistinct white mass."

When the skies again cleared on January 7, the party could see that the "white mass" was the northern point (Cape Selborne) of a highland that continued south, veering slightly to the east. North of this was another wide gap before the land that had been sighted several days before (Britannia Range). Barne knew that the southern party would have seen the distant mountains on its journey south, so there was no purpose in his exploring farther in that direction. He now set his course to the broad gap that had appeared that day, in hopes of traveling up the divide to see whether its ice surface continued level or rose as a glacier toward the interior.

The party was hardly under way when another blizzard came down, pinning it in camp. This was followed by fog with no visibility or surface definition. Out of frustration on January 12, the party pulled in the fog following the compass bearing, but several times one of the men fell in a crevasse, so they halted again. The bad weather held them until January 15, the day their rations dictated that they turn back. As they returned along their old tracks, the question of archipelago or continent remained unanswered.

To be sure, Armitage's western party had found that Ferrar Glacier drained from an ice plateau, but their traverse onto it had been less than twenty miles. If the major gaps seen by Scott's and Barne's parties were in fact straits that passed through to a frozen sea on the far side, the ice cap discovered by Armitage's party might drop off to the west and to the south to a shoreline bounded by the same floating ice sheet. With *Discovery* frozen tight in Winter Quarters Bay for a second season, Scott's expedition was given another opportunity to answer the question.

For the austral summer of 1903–1904 (see Chapter 2), Scott chose to lead the western party onto the plateau following Ferrar Glacier. Armitage led a party exploring the upper reaches of the Blue and Koettlitz Glaciers in an attempt to find another route through the mountains, but he did not. And Barne was again sent to the southwest in an attempt to determine the nature of the gap between the Britannia Range on the north and Cape Selborne on the south.

The southwest party consisted of twelve men, six of whom were in support and would turn back one week south of the Bluff depot. They pulled out of Winter Quarters Bay on October 6, arriving at the depot two weeks later. In locating the depot, the party made an unexpected discovery of considerable importance: the depot was no longer on the alignment between the summit of Mount Discovery and the cone at the end of Minna Bluff but rather had moved 450 paces toward the north since its placement the previous year. The ice shelf was actually moving toward the open sea. On the return leg a careful survey by Mulock of the depot and Mount Discovery showed that the ice shelf had moved a total of 608 yards in 13½ months, or at a rate of approximately five feet per day. This was the mechanism that replenished the Barrier when great icebergs rifted from its margin. What could the origin of such a massive movement be? Where was the head for such a flow?

After a good start with sails on October 21, the party had much tougher pulling south of the depot on the following two days. On the evening of the 23rd a fierce blizzard struck, holding the men in their tents until the 28th, the day that the support party needed to return. Prevented from carrying supplies farther along for the southwest party, and thereby compromising perhaps a week of exploration, the men of the support party headed north with the gale at their backs. When the storm had not abated by afternoon, Barne gave the order to head into the teeth of the wind, six men hauling three sledges in train, steering by compass as they moved forward. The following day offered the only fine weather in the next two weeks. With only occasional moments of clearing, snow storms and fog plagued the party most of the time as they crept their way to the south-

Figure 4.7. Barne's route to the mouth of Byrd Glacier failed to reach an outcrop of rock due to the intense crevassing where the glacier enters the ice shelf (see Fig. 4.2).

west. On November 7 they spotted Cape Selborne, and to their right (west) a dark ice-free shoulder with what they took to be volcanic craters. (This area to the north of the mouth of Darwin Glacier is actually composed of granite, but the name of the feature remains Goorkha Craters.)

On November 8, still about fifty miles to the northeast of the mouth of the strait, the party began to encounter broad undulations in the ice surface spaced about three hundred yards apart, arching out from the opening. Over the next several days these features grew in amplitude and became increasingly more crevassed. On the 13th Barne thought it prudent that the party rope up. The undulations now pointed toward the strait, and the party followed parallel to them, about two miles to the south of a zone of increasingly disturbed crevassing that extended back to the cape at the mouth of the strait (Fig. 4.7).

The men covered twenty-five miles over the next four days, hauling over and around increasingly steeper and higher ice ridges. By the evening of November 17 they were abreast of Cape Selborne to the south and about four miles from rock outcrop at the northern portal. From the top of a high ice ridge Barne looked west into the strait. Both the north and south sides of the channel ended in bluffs, and between them was a distinct horizon of level ice. (The bluff seen on the south side was Mount Tuatara, located about midway down Byrd Glacier, and the bluff on the north side was the western end of the Britannia Range [see Fig. 4.2].) Mulock's measurement of the angle of rise of the horizon, a mere one thirtieth of one degree, was so slight that doubt remained as to whether the horizon was on an inland ice shelf or actually on an ice sheet. Nevertheless, all the evidence of ice ridges and crevassing led Barne and Mulock to conclude that they were standing on a glacier that was flowing toward them from the interior.

The last objective then was to collect a piece of the rock from the cliffs to the north of the glacier. Given the curve of the bluff, Barne could see that moving camp several miles to the west would bring the outcrop closer, but before the party could move three miles upstream it was pinned by a blizzard for another thirty-six hours. With improved visibil-

Figure 4.8. Viewed along the axis of the chasm from west to east, this is the chaos of crevasses that prevented Barne's party from reaching bedrock.

ity on November 20, Barne sketched and Mulock surveyed the mountains to the north and west. In the afternoon the party pulled toward the cliffs, but within a mile the glacier surface had become so rough that it jeopardized the runners on the sledges, and the men retreated to the east.

Two days later they tried to reach rock again (see Fig. 4.7). This time roped up and without the sledges, the party headed into some of the most savagely crevassed ice on the continent (Fig. 4.8). This part of the glacier was blue, with at least 50 percent of its surface regaled with stubby, lens-shaped crevasses both open and bridged. These were relatively easy to negotiate, but longer and wider crevasses also crossed the area, requiring careful probing and many detours. At places it was necessary to climb down into an open crevasse in order to cross it. After about three miles of maneuvering an increasingly hazardous landscape, the men were stopped by a chasm that dropped off perhaps eighty feet and was filled in its depths by a jumble of seracs. It was as if the glacier had opened from below, and into the abyss had fallen the surrounding ice.

The feature that Barne's party witnessed was similar to what Scott's party had encountered in its attempts to reach Cape Selborne and Cape Wilson. Where the outlet glaciers of the Transantarctic Mountains enter the Ross Ice Shelf, their movements make great tears along the marginal interfaces, with openings that run deep into the floating ice.

Thwarted by this frozen moat, the party returned to camp, but the glacier offered a consolation prize in the form of a medial moraine with a sparse line of boulders that was streaming down from one of the rocky points to the west. The rock was granite, both pink and gray, as had been found at Granite Harbour. This was the proof of the continuing continentality of these rocks with those in the McMurdo Sound area. Ferrar would be a happy geologist when he saw these samples and compared them to the granites that he had collected that season in the lower Ferrar Glacier drainage.

Barne had also wanted to approach the land to the north of the Goorkha Craters (Cape Murray) on his return, but it took the party too long to round the eastern end of the disturbed zone. November 25 was a clear day when Barne sat down and executed a sketch of the coastline that went from the southern flank of Mount Discovery all the way to Cape Selborne and covered sixteen running feet of paper. In concert Mulock carefully ran sights and angles on all the peaks.

Over the next week overcast skies with intermittent storms forced the men to navigate mainly by compass as they unknowingly veered west of the line back to the Bluff depot. When clouds lifted enough to see landmarks along the western coast, they were able to set their course right, and they were then aided by a strong southerly breeze that filled their sails, landing them at the depot on December 5. A rock collected for Ferrar from the end of Minna Bluff was the last objective of the party, which then pulled on in to the ship on December 13, having covered the final distance in six days.

In addition, to its considerable discoveries in Victoria Land, the *Discovery* Expedition extended the known length of the Transantarctic Mountains another 250 miles to the south, with no end in sight. In so doing, the expedition set a record for the highest latitude reached at either pole. Due to the limited views into Barne and Shackleton Inlets, however, uncertainty remained as to whether the mountains south of Barne Inlet constituted a continuation of the continent or were an extended archipelago of islands. The answer would be left to Shackleton and the next expedition to assault the pole.

In promoting the British Antarctic Expedition of 1907–1909, Shackleton had been explicit that two goals would be reaching both the Geographic and Magnetic South Poles. In the spring of 1908 a number of short sledging trips paved the way for the southern party, including the staging of supplies at the old winter's quarters at Hut Point. A depot party cached 167 pounds of pony food (four days' supply) and a gallon of oil 138 miles south at Depot A. The northern party, David, Mawson, and Mackay, left Cape Royds bound for the Magnetic South Pole on October 5.

On October 29 the southern party, Shackleton, Adams, Marshall, and Wild, set out on their campaign, with four ponies, the motor car, and a support party of five. The motor car turned back before reaching Hut Point, where the combined parties assembled for several days before heading south across the ice shelf. Thirty-eight miles south of Hut Point the support party turned back, with hearty farewells and good luck to the four whom they left.

With their four ponies each pulling an eleven-foot sledge weighing six hundred pounds, the southern party faced a distance of 1,450 miles to the pole and back again to Cape Royds. Before they had gone a mile, a storm forced them to pitch camp. They were laid up the following day as well, but on November 9 the winds were calm and the sun shone brightly. Crevasses were an immediate problem, but these were left behind as the party rounded Minna Bluff, setting a course due south across the ice shelf.

The four ponies, named Grisi, Socks, Quan, and Chinaman, pulled fairly strongly, but their small hoofs sank deep into the snow, sometimes up to their bellies, so progress was hard won. Mileages were a respectable fourteen to sixteen miles per day, but already it appeared that food would be a critical factor. The weather was generally good as the party crossed the ice shelf, its surface "as wayward and changeful as the sea," and the men's spirits were high.

The party reached Depot A on November 15. The first of the ponies to play out was Chinaman, which was shot on November 21. The horsemeat was an important part of the men's diet, with a portion carried along and the remainder depoted with the carcass for the inward march. Probably thanks to this supplement of fresh meat, none of the party suffered from scurvy throughout the journey, although dysentery became a worrisome problem on the return leg.

On November 22 the party spotted new land beyond Mount Longstaff, the southernmost sighting of Scott's party in 1902. Because they were farther out on the ice shelf than Scott, the men of Shackleton's party were also able to see high peaks beyond the foothills to the south of the strait. For the next four days the party pulled toward that unmarked line in the snow touched by the toes of Scott, Wilson, and Shackleton five years before.

On November 26 the party crossed Scott's "farthest south," the men congratulating themselves with a four-ounce bottle of curaçao and smokes as they turned in for the night. During that day the mouth of the strait had opened wide to the west. On the north side of the channel a new range with high summits appeared running northward, which Shackleton guessed connected through to Mount Albert Markham farther to the north. On the south side of the strait, a long, intricate ridge system branched down to the west from the summit of Mount Markham. And as far in the distance as one could see there were vague mountains, more of them right out to the horizon. Surely this was a major outlet glacier, but a straight line at the horizon like Barne and Mulock had seen was not so sure between the distant blocks. On his map of the expedition's surveys, Shackleton showed only "Shackleton Inlet" for this breach in the mountains, although he soon would know that the ice sheet that Scott had penetrated for 150 miles to the west of Ferrar Glacier continued along the backside of the western mountains to this southerly latitude and on to the pole.

Given that the new mountains appearing to the south were arching toward the southeast, it became clear that the chain would eventually cross the meridian that the party was following to the pole. Consequently, the men altered their course to the east of south, hoping still for an end run. During this stretch of the journey, Shackleton noted, "It falls to the lot of few men to view land not seen by human eyes, and it was with feelings of keen curiosity, not unmingled with awe, that we watched the new mountains rise from the great unknown that lay ahead of us."

As the men passed Mount Longstaff, they could see that it was actually an elongate range trending north-south that had appeared as a solitary summit when the mountain was viewed end on and from a distance. Now the range opened to their right. A series of graceful spurs rose steeply to a ridgeline punctuated by a row of peaks rising to between nine thousand and ten thousand feet. Narrow, heavily crevassed glaciers dropped between the spurs, merging into smoothness at the foot of the range. The distance from the coast to the crestline of these mountains was only about twelve miles. Nowhere between

Mount Melbourne and here did such a lofty range rise directly up from the shore. Typically the coast was mantled by a low piedmont or by relatively diminutive ranges, with the highest peaks set well back into the mountains. In northern Victoria Land, of course, the highest mountains ascended directly from the sea, but for the next six hundred miles down to the Longstaff Peaks, as they came to be called, this pattern did not exist. (Nor do such high mountains occur on the coast throughout the remainder of the Transantarctic Mountains. The Longstaff Peaks, and the Holland Range to which they belong, indeed represent a physiographic anomaly.)

Marshall surveyed this majestic range on November 28, from a latitude of 82° 38' S. That night Grisi was shot for the next depot (Depot C). With twelve hundred pounds of provisions for nine weeks of labor, the party set out the next morning planning on two men helping each of the two remaining ponies pull the sledges. But the ponies stopped pulling when the men were in their traces, so they unhooked and let the animals pull alone. On that day more mountains rose behind the Longstaff Peaks, high, white, blocky, estimated to be ten thousand to fifteen thousand feet in height. By the 30th more mountains were rising farther to the east of south and the party faced the reality that they would have to search for a passage through the mountains if they were to reach their goal.

They had been on the trail now thirty-three days, the pulling had been hard, and already a great hunger gripped the party, but surface conditions had been better and progress had been significant compared with the traverse in 1902–1903 on which Scott, Wilson, and Shackleton had been beset by a mist of ice crystals and near-melting temperatures that made pulling almost an impossibility. Even as Shackleton had turned back in 1902, he had recognized the possibility that the Western Mountains might continue for hundreds or even thousands of miles in their trend to the south-southeast, and that any assault on the pole from the Barrier was likely to involve crossing these mountains. Armitage had shown that it was possible. But still, if the mountains had terminated, and the Barrier merged into the ice sheet around the end of the last mountain, such a route would surely have been better than a route through an unknown, crevassed corridor of the mountains. The time had come for Shackleton's party to face its destiny.

On December 1 the party began to climb across broad, shallow undulations on the Barrier. That evening Quan was shot and depoted. The next morning the four men took one sledge, Socks took the other, and they dragged toward what appeared to be a broad inlet beyond a headland of rock at the coast. In the afternoon a huge array of pressure ridges and crevasses appeared ahead of the party, extending from the far side of the headland across the path and eastward out of sight. The smoothest ice was toward the reddish hill that stood just to the near side of the disturbance, so the men headed straight for that, camping about eight miles off, with the intention of climbing to its two thousand–foot summit for a view to the south.

Thoughts of what they would find ran through the men's minds as they recorded their diaries and fell to sleep. The pressure field at the mouth of the inlet was similar to ones at Barne Inlet and Shackleton Inlet, suggesting that a major glacier might be flowing through the mountains here. If so, what would its condition be? Would it plunge steeply or flow smooth? Would crevasse fields of impossible complexity bar the way? They could only hope that a route would be clear, and so they named their summit Mount Hope.

116

Figure 4.9. Flanked by crevasses, Mount Hope stands
at the mouth of Beard-
more Glacier. Shackleton's
reconnaissance route to
the summit is shown in
orange, and his traverse
through The Gateway and
onto Beardmore Glacier is
shown in yellow. The dots
locate campsites in The
Gateway and in front of
Granite Pillars.

The men were up at 5:30 A.M. Socks was tethered and given a day's ration of maize, the camp was left standing, and the men headed off with a biscuit lunch in their pockets. After the first couple of subtly bridged crevasses the men roped up and continued into an increasingly crevassed terrain that culminated in a chasm eighty feet wide and three hundred feet deep, similar to what Shackleton had seen at Cape Wilson, but larger. Fortunately, as the men followed the edge of this tear to the east, it narrowed and finally pinched out in a snow-covered bridge, which they crossed safely. From there they crossed more crevasses and several pressure ridges before arriving at smooth blue ice about 12:30 P.M. Here the men refreshed themselves with a couple of biscuits and meltwater formed at the side of a large granite boulder embedded in the ice, then quickly crossed to the outcrop of bedrock, and recorded the southernmost landfall on Earth to date (Fig. 4.9).

The rock was granite with a reddish coloration probably owing to its weathering. The route up the mountain followed the crestline of the elongate massif. Because of the crumbliness of the granite, finding good footholds was a challenge, and on a number of occasions one of the men sent a loose boulder tumbling down the slope. Because of the rounded surface of Mount Hope, Shackleton speculated, correctly, that it had been overridden by ice at a previous time, when it stood higher than today. The rocky lower slope

Penetrating the Interior

Figure 4.10. View from the summit of Mount Hope up Beardmore Glacier. The bottom panel is a photograph taken by Shackleton's party in December 1908. The upper panel is a rephotograph taken from precisely the same spot at the summit of Mount Hope in December 2010. Mount Kyffin is the sharp horn on the left skyline; Wedge Face is the dark buttress immediately to its right; The Cloudmaker is at the center skyline; and Granite Pillars is the rocky buttress on the right side that casts a long shadow. Note that touch-up omits Wedge Face on the lower panel (see Fig. 4.11). Rephoto by John Stone.

gave way to a snow surface higher up, followed by a short rocky stretch to the ridgeline. In a flash, apprehension melted into deliverance, as the party topped the ridge and looked upon a great causeway furrowing through the mountains on a direct line to the pole (Fig. 4.10).

After a short scramble to the summit, the men surveyed their new domain. The height of the mountain was measured by both aneroid and hypsometer at 3,350 feet (although today's maps gauge it to be about 2,740 feet). The glacier appeared to extend far inland and to merge there with high, inland ice. At the foot of the massif, where the glacier entered the ice shelf, great pressure ridges rose up and deep ruptures extended for many miles into the ice shelf. But upstream, at least from this vantage, the glacier seemed relatively smooth and crevasse free, especially along the western margin.

From this high vantage the mountain front continued even farther to the southeast, Shackleton speculated to 86° S, dissolving into the Barrier at the horizon. On either side of the glacier, ridge systems rose abruptly, with those to the east largely bare of ice and those to the west largely glaciated. In the distance, maybe sixty miles up the glacier, the mountains formed bluffs, with the appearance of horizontal stratification, as had now been seen at numerous places throughout the Western Mountains. And out beyond that, vague semblances of even more mountains strained the eye to see.

However, that day the most intriguing discovery was on the west side of the glacier maybe fifty miles away. Shackleton relates, "In the far distance there is what looked like an active volcano. There is a big mountain with a cloud on the top, bearing all the appearance of steam from an active cone. It would be very interesting to find an active volcano so far south. After taking bearings of the trend of the mountains, Barrier and glacier, we ate our frugal lunch and wished for more, and then descended."

The descent began at 4:00 P.M. Following back the trail they had pieced together through the crevasses, the party reached camp three hours later. Shackleton was again snow-blind because he had taken off his goggles early in the day in the crevasses and hadn't put them back on. Time and rations were short, but Shackleton could calculate success in reaching the pole if the glacier surface was kind and the weather held. The course was set. A gateway to the glacier had been spotted the day before from the summit of Mount Hope, a narrow path of white that cut behind the mountain where the party could avoid the pressured zone at the mouth of the glacier.

They were off by 8:00 A.M., with Shackleton, Marshall, and Adams pulling one sledge and Wild tending Socks with the other. In the event that the man-haulers found a crevasse, they would be able to warn the driver with the pony. It took until well into the afternoon negotiating crevasses before the party reached the foot of the snow slope that they planned to take across to the glacier. This proved to be much longer and steeper than they had judged from the summit, but by 5:00 P.M. they had reached the pass. Shackleton described his impressions of that moment: "The pass through which we have come is flanked by great granite pillars at least 2000 ft. in height and making a magnificent entrance to the 'Highway to the South' [Figs. 4.11, 4.12]. It is all so interesting and everything is on such a vast scale that one cannot describe it well. We four are seeing these great designs and the play of nature in her grandest moods for the first time, and possibly they may never be seen by man again." Shackleton could not have imagined the International Geophysical Year fifty years later.

From its camp a little south of the pass, the party moved farther south on December 5. Almost as soon as the men reached the glacier they were enmeshed in a thicket of small crevasses in blue ice. The main danger was that Socks would stumble in one of these narrow slots and break his leg, so the three men shuttled the two sledges for much of the time while Wild carefully led Socks through the maze. When a large swell of pressured blue ice rose up, the party found passage at the snow-patched margin of the glacier next to a set of towering granite pillars, where they camped for the night. At this site they made a depot (Lower Glacier Depot or Depot D), leaving food enough for six days.

The following day the party crossed another half-mile of crevassed blue ice, with the trio shuttling the sledges and Wild bring up Socks, but from there on they traveled along the snowy margin of the glacier with a pressure swell to the left and retreating valleys to the east. The routine returned to the three man-haulers spotting crevasses and Wild and the pony following close behind. The next day the routine was the same. Crevasses appeared along the passage, trending in from either side, but they were avoided without problem.

Then shortly after the lunch camp, disaster struck. Socks broke through a crevasse bridge that the man-hauling team had unknowingly crossed. The sledge lurched forward, buckled down into the crevasse, but held. Wild was jerked forward into the crevasse by the lead line before catching himself with his left arm on the far side. The others quickly pulled him out and stabilized the sledge. Fortunately for the survival of the party, Socks's weight had snapped the swingle tree of the sledge as the pony disappeared into the "black bottomless pit." Had it not, the sledge would have gone down with the pony, and with them two of the four sleeping bags carried by the party, without which they probably would not have made it back to Cape Royds. The loss of Socks, however, was critical, for he was to have been the foodstuff that would carry the party to the pole. The pony's maize became the meager substitute for his calories lost to the crevasse.

From here onward the four men pulled the two sledges in train when possible, and relayed them when the surface was too rough or slick. Crevasses became a common part of their routine. You fell in one on your harness, the others helped you out, and you went on, watchful as could be. Everyone's shins were bruised from catching them on the lips of blue ice when a foot went through a bridge. A great blessing was the weather, which remained clear and windless, allowing daily progress up the glacier.

On December 10 the party had pulled abreast of the massif, which from Mount Hope had appeared to be a volcano. Now in under the cloud that had hung over the massif since starting up the glacier, it was clear that the mountain was no volcano, though it certainly had a knack of holding a cloud at its summit, and thus its name, The Cloudmaker. In form, it was a bare face of rock about twelve miles long, steeply rising to a crest more than three thousand feet above the side of the glacier, with no spurs patterning its bulging wall. After dinner Shackleton climbed up the mountainside while the others ground maize into meal with rocks chosen from the moraine at the edge of the glacier. For the first six hundred feet or so up the face, the bedrock continued to be covered in moraine, left by the glacier at some time in the past when it stood much higher in this valley. Above that Shackleton found outcrop. The rocks were slates and sandstones, and a brown rock that he could not identify. The party had no geologist, but Shackleton collected small samples for identification by those at Cape Royds.

From his vantage Shackleton peered to the south. Low cumulus clouds hung in the upper reaches of the glacier, as they had for a number of days now, so the distance to the plateau remained obscure. The mountains to the east, on the far side of the glacier, formed a sort of amphitheater, thirty miles across, rising through broad slopes of white to a balcony of summits above eight thousand feet. At the left (northern) end of this arc, a ridge of bedrock asserted itself as a bold promontory into the glacier. Its triangular, faceted face reminded Shackleton of the photos of Cathedral Rocks on Ferrar Glacier, and Wedge Face was the name he gave to this feature.

Scattered across the lateral moraine at the foot of The Cloudmaker were conspicuous boulders of limestone breccia, a rock composed of angular fragments of limestone stuck in a matrix of calcite "showing a great mass of wonderful colors." Several small samples of this limestone were also collected for identification by the geologists.

For the next several days the party pulled south past The Cloudmaker (Fig. 4.13). On December 14 they camped at 5,600 feet, about fifteen miles south of the misty massif, with hopes that they would soon be on the plateau. Mountains had been rising to the southwest for several days, but the grand spectacle was a grouping of summits and promontories that emerged to the west of The Cloudmaker and were now arrayed to the north and northwest of camp around a high cirque whose glacier dropped abruptly through a breach in the bedrock bounding the main glacier. The crest of the cirque and the ridgeline

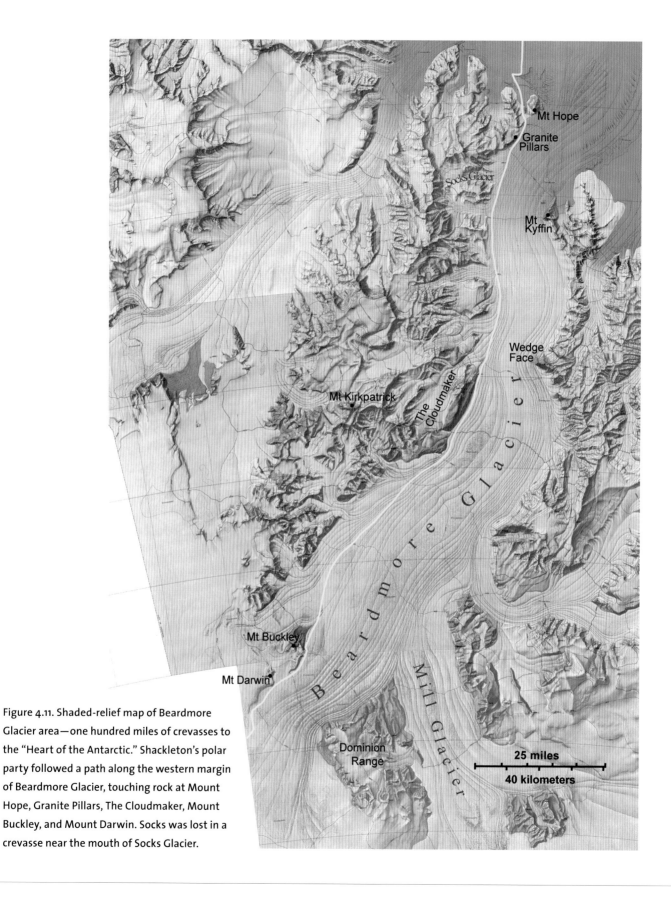

Figure 4.11. Shaded-relief map of Beardmore Glacier area—one hundred miles of crevasses to the "Heart of the Antarctic." Shackleton's polar party followed a path along the western margin of Beardmore Glacier, touching rock at Mount Hope, Granite Pillars, The Cloudmaker, Mount Buckley, and Mount Darwin. Socks was lost in a crevasse near the mouth of Socks Glacier.

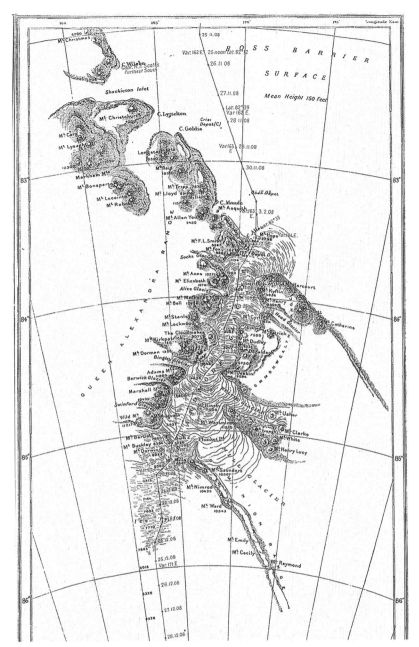

Figure 4.12. Shackleton mapped the mountains surrounding Beardmore Glacier and his route along it. Note the side route to Mount Hope from the "3.2.08" [*sic*] campsite and compare with Fig. 4.9. By all accounts Shackleton's party held close to the western margin of Beardmore Glacier for most of the ascent. This map depicts the campsites as close to midstream, probably license taken by the cartographer, Eric Marshall, who was not a member of the polar party and worked from their sightings.

that ran to the east of it were of a remarkable height. The highest summit at the west side of the cirque was surveyed by Marshall at 14,624 feet. Named Mount Kirkpatrick for one of the early supporters of the expedition, this is now known to be the highest point in the Transantarctic Mountains, with an elevation of 14,856 feet.

The camp on December 16 was a few miles downstream of a three-peaked island that split the western side of the glacier. There the party planned to depot four days' worth of food and any extra clothing that they were not wearing, to make their final dash for the pole. It seemed that here at last the glacier would meld into the plateau. The mountains

Figure 4.13. A broad piedmont glacier puddles on the south flank of The Cloudmaker. Shackleton's party rounded the promontory on December 13, 1908, and headed south along the band of blue. Wedge Face stands above the clouds at the right rear of the image.

formed long bluffs to both the east and west of the glacier, in places with narrow, tributary glaciers spilling over their fronts. The bluffs on the east side of the glacier were also separated by wide tributaries to the main glacier. Horizontal stratification was visible in all of these mountains, though they were too distant for the men to be sure of the rock types. Shackleton commented, "These mountains are not beautiful in the ordinary acceptance of the term, but they are magnificent in their stern and rugged grandeur."

The next day was a difficult ascent over blue ice as they passed the nunatak (Buckley Island) where, in the afternoon, they made their depot, leaving all but the barest necessities (Fig. 4.14). Toward the end of the day they progressed by climbing the steep blue ice to the distance of an alpine rope, cutting a shelf in the ice where the four could stand, and then pulling the sledges up one at a time to that point, repeating the operation until they reached a level spot where the tent could be pitched, even though there was no snow to hold down the flaps.

After dinner Wild climbed to the top of the nunatak and came down reporting that the plateau was at hand beyond the hill. This was good news indeed, for it meant that soon the men would be free of crevasses and blue ice. However, Wild had also discovered that the layers of sandstone in the upper part of the nunatak contained six or so beds of coal, ranging in thickness from four inches to eight feet. Compared to the wisps of car-

Figure 4.14. Mill Glacier merges with Beardmore Glacier in front of the Dominion Range, left rear. With the central portion of Beardmore Glacier heavily crevassed, Shackleton chose a more snowy route along its margin. Buckley Island and Mount Darwin, the party's last landfalls before the polar plateau, bound Beardmore Glacier in its headreaches, right rear.

bon that Ferrar had found in the McMurdo area, this was a major deposit and would be one of the most important geological discoveries of the expedition.

The southern party reached the plateau on December 18. At lunch they camped near the last point of land in the headreach of the glacier, a solitary peak surrounded by ice, which they named Mount Darwin (see Fig. 4.14). The rock cropping out there was different from the sandstones on Buckley Island. Shackleton walked over to the outcrop while lunch was being prepared and found a gray, layered rock that was later identified as limestone. He collected two samples, which would be the trophies of the expedition, these pieces of the "southernmost rock," as rare as moon rock to a later generation.

The day was a struggle over rising blue ice with the men alternating between pulling in harness and relaying with the alpine rope. They carried five weeks' worth of food and were about three hundred miles out from the pole, or about six hundred miles from their last depot if they made the distance. By the time they bedded down for the night at seventy-four hundred feet, they still had not left crevasses behind. Nor would this situation change for another week. What did turn was the weather, with a succession of hard, steady winds blowing out of the south directly into their faces.

Excerpts from Shackleton's diary show the hope and frustration that the plateau was inflicting:

December 19.—Not on the plateau level yet, though to-night 7888ft. up, and still there is another rise ahead of us.

December 20.—Not up yet, but nearly so.

December 21.—We have been hauling the sledges up, one after another, by standing pulls across crevasses and over great pressure ridges.

December 22.—All day long, from 7 A.M. except for the hour we stopped for lunch, we have been relaying the sledges over the pressure mounds and across crevasses. Our total distance to the good for the whole day was only 4 miles southward, but this evening our prospects look brighter, for we must now have come to the end of the great glacier. It is flattening out, and except for crevasses there will not be much trouble in hauling in sledges tomorrow.

December 23.—Eight thousand eight hundred and twenty feet up, and still steering upwards amid great waves of pressure and ice falls, for our plateau, after a good morning's march, began to rise in higher ridges, so that it really was not the plateau at all.

December 24.—A much better day for us; indeed, the brightest we have had since entering our Southern Gateway.

Early in the day the surface had begun to smooth out and by the time they camped, the party felt that they had left the crevasses behind. They also were buoyed by thoughts of Christmas Day, when all the food plans of the previous month would be realized in holiday feasting.

On a better than average breakfast the men marched on Christmas Day as always. Soft snow was drifting from the south, placing extra drag on the sledges, and the surface continued to rise steeply. By the time they camped at 6:00 P.M., the air had clarified in the southeast quadrant and new lands could be seen stretching far into the distance in that direction.

The Christmas dinner was bolstered by small fractions of food that had been saved by each cook over the previous four weeks. Goodies included some of the emergency Oxo beef stock and a small plum pudding boiled in cocoa with a spot of medicinal brandy. The feast ended with "cigars and a spoonful of *creme de menthe* sent us by a friend from Scotland," and the discussion shifted to further reduced rations if the party was to succeed in reaching the pole.

Famished but fit, with no signs of scurvy, they continued south. By December 28 they were camped above ten thousand feet. They had crossed their last crevasse the day before, but the surface continued to rise. The altitude was taking its effect, especially on Shackleton, who was experiencing severe headaches. A storm sprang up about noon on the 30th and held them in their tents the rest of the day. By January 3 the men finally admitted the impossibility of reaching the pole, if they wanted to return alive. At that camp they made a depot of food and fuel to see them back to the depot at the head of the glacier, and with one tent and ten days' food made their final (and futile) rush toward the pole. Even with the lightened load, the "rush" was held to ten to fifteen miles per day by blowing snow, thin air, temperatures 20–30° F below zero, and the meager ration. On January 7 "a blinding, shrieking blizzard" descended on the party, holding the men in their tent through the following day. Miserably cold and cramped, they turned out on January 9 for what was to be the last day of the outward march. Without the weight of the sledge they half-walked, half-ran for five hours due south. At 9:00 A.M. they stopped, sighted their location, raised the Union Jack, took possession of the plateau in the name of King Edward VII, and recorded a photo. Shackleton wrote of the moment, "We have shot our bolt, and the tale is latitude 88° 23′ South, longitude 162° East."

The inward march was bound to be a tenuous affair. The famished party would need to average fifteen to twenty miles per day or run out of food before reaching each of the six depots that had been laid on the outward journey. They would first need to find each of the six depots. They would need to march even if weather was bad. From here on Death would stalk the party's every step, draining energy and body heat from the desperate four.

But Providence also guided their course. On the plateau the headwinds that had been so cruel on the outward march now aided the return, as the men rigged a sail from the tent floor and propelled the sledge along. Abandoning all caution, the weary foursome plowed over crevasses without testing bridges, even though each of them had his share of freefalls into the slots. Somehow they managed to make their distances. On January 11 they were back to their last depot (Depot F), on the 17th they reached the Christmas depot, on the 20th, the depot at Buckley Island at the head of the glacier. There they loaded up the food, the second tent, the extra clothing, and samples of rock.

On the traverse down the glacier the party ran out of food the day before reaching the Lower Glacier depot by Granite Pillars. They were so exhausted that they collapsed two miles short of it, and Marshall hiked the last distance to bring back a meal. Three times he fell into a crevasse, but he caught himself each time with his arms. Revived by the nourishment, the party moved on to the depot. Here too Wild collected a final sample of limestone breccia on the moraine at Granite Pillars, and the men loaded their supplies for the next leg, fifty miles to Grisi depot on six days' rations. Another hardship now was added to the burden of the party. First Wild developed dysentery, then Shackleton too, probably from the horse meat. They reached Grisi depot essentially out of food on February 2 and had a filling meal. The following day the entire party was wracked with dysentery but marched anyway. The next several days were miserable, but eventually the diarrhea ran its course and the men's health improved. Aided by a strong following wind and their sail, the party reached Chinaman depot on February 13, again with no remaining rations. A good meal, including the pony's liver, and the party moved on. On February 20 the men reached Depot A, the one laid by Shackleton's party in the early spring. Here were three days' rations to see them the distance to the Bluff depot, where Joyce's party was supposed to have left ample supplies for the last leg to Cape Royds. If they made it to the Bluff, they knew they would survive.

On February 21 the party pulled twenty miles in a blizzard. The following day the weather cleared and their spirits surged as they crossed the tracks of four men with dogs. These had to have been from Joyce's depot party. The next day the southern party set off at 6:45 A.M. At 11:00 A.M. Wild caught sight of flags on the depot reflected in a mirage. The men immediately devoured their remaining few biscuits and hurried toward the flags. However, the distance was deceiving because the flags were on bamboo poles atop a ten-foot high mound of snow, and the men struggled another five hours until they at last reached the altar of food. Shackleton scrambled to the top of the snow mound and dug out tins and boxes "containing luxuries of every description," which he tossed down to his giddy mates. They had survived. Would food ever taste so good again?

The geographical discoveries of the British Antarctic Expedition of 1907–1909 were considerable. The vastness of the inland ice was measured by the traverses of the northern and southern parties in their quests for the two South Poles. These lines on the map flanked a similar line drawn by Scott's western party in 1903. Together they showed that an ice sheet of ever rising elevation extended several hundred miles into the interior of the continent. Because the elevations kept rising, the crest of this great sheet of ice had to be even higher in the interior. The Western Mountains, first sighted at Cape Adare by Ross's expedition, were now known to extend for more than one thousand miles, and throughout that length to be the barrier between the inland ice sheet and the Ross Sea and Ice Shelf. There could be no more question about the mountains being an archipelago of islands. Without a doubt, they formed a continuous mountain belt which dammed the ice of the plateau. The straits that crossed this belt were outlets for the inland ice, great glaciers flowing inexorably downward through the mountains, tearing gigantic gashes in the ice shelf where they met.

Shackleton's southern party also made geological discoveries that were of considerable importance. When the limestone breccia from the Lower Glacier depot was examined back in Melbourne, it was found to contain fossils of a sedentary, cone-shaped invertebrate called *Archaeocyatha*. These diminutive filter feeders with large pores connecting their inner and outer walls date from the Lower Cambrian, the oldest fossiliferous horizon on the geologic timescale. Several years later, Griffith Taylor, the geologist on Scott's *Terra Nova* Expedition, was able to identify several different species, with the largest complete specimens measuring no more than a third of an inch. The breccia from the Lower Glacier depot also yielded tiny fossils of branching, calcareous algae called *Epiphyton*, as well as sponge spicules. One of the samples of limestone from outcrop at Mount Darwin also contained tiny fragments of *Archaeocyatha*. Although it was too scrappy for specific identification, it nevertheless was the first *in situ* fossil of Cambrian age found on the continent, and it remained the only one for another fifty years.

5 Beyond the Horizon
Discoveries in the Queen Maud Mountains

Scott's Final March

From the planning stages of the British Antarctic Expedition of 1910, Scott knew that he would be following largely in the footsteps of Shackleton. The winter quarters would be on Ross Island, as deep into McMurdo Sound as ice conditions allowed. Scott had claimed that real estate nine years before. And from Minna Bluff to the mouth of Shackleton Inlet, the polar party would be on the old trail that Shackleton, Wilson, and he had blazed in 1902. South of that was Shackleton territory, with its proven route through the mountains at Beardmore Glacier and onto the polar plateau. Only after he had closed on the last 111 miles to the pole would Scott be breaking a new trail. To be sure, the expedition had been touted for its scientific goals, and the learned societies had all stepped forward in its support, but clearly to the British public the South Pole was the prize. Given the discoveries of the two prior British expeditions that had opened the gateway to the interior, was it not the British destiny to set foot first at the pole? And was not Robert Falcon Scott the man to do it?

The expedition sailed from London on June 1, 1910, aboard the *Terra Nova,* already a veteran of Antarctic waters as the relief ship second to *Morning* in the austral summer of 1903–1904. When the ship arrived in Melbourne on October 12, Scott was shocked to receive a cablegram from the island of Madeira that read, "Beg leave to inform you *Fram* proceeding Antarctic. Amundsen." For several years Roald Amundsen had been garnering support for an expedition of his own to sail the *Fram* into the Arctic Sea, drift with the ice for several years conducting oceanographic research, and perhaps take a side traverse with the goal of reaching the North Pole. News had broken the first week of September 1909 that both Cook and Peary claimed to have taken that prize. Apparently Amundsen had somehow redirected his Arctic voyage to the south. Could there be any doubt that his goal was now the South Pole?

Here was a daunting rival to Scott. Rumored American, Japanese, and even German expeditions that might be targeting the pole could be dismissed as amateurish or ill equipped. Not so the Norwegians. Roald Amundsen was renowned for having recently sailed the fabled Northwest Passage, the quest of three centuries of Arctic exploration. At the age of twenty-five, he had been among the first men to winter over in Antarctica, while serving as first mate aboard the *Belgica,* an ill-equipped Belgian exploring ship. Commanded by Adrien de Gerlache, the *Belgica* had frozen fast to the west of the Antarctic Peninsula in March of 1898 and drifted with the ice until breaking free the following March.

What the Norwegians were up to was a nagging question in the minds of the members of the *Terra Nova* Expedition as they established their base on Ross Island. They chose Cape Evans, about midway between Cape Royds and the old winter quarters at Hut Point (see Figs. 1.17, 2.7). Any existing doubts vanished on February 4, when *Terra Nova,* returning from a cruise toward King Edward VII Land, found the *Fram* moored in the Bay of Whales at the eastern side of the Barrier. Each party showed the other the utmost courtesy. The officers of the *Terra Nova* were invited to Framheim, the expedition's quarters built several miles in from the ice edge, and Amundsen and two others were received on board the *Terra Nova.* The intelligence that Lieutenant Campbell reported to Scott was that Amundsen headed a ground party of nine, and that they were well equipped and had more than one hundred dogs.

So the race was set for the austral summer of 1911–1912—the Norwegians with their dogs, and the British with motor-sledges, ponies, a few dogs, and man-hauling. Furthermore, the Norwegians' location at the Bay of Whales provided them with a starting point sixty miles closer to the pole than the British had at Cape Evans. For Scott there was nothing to do but stick to the game plan, lay the depots in the fall and early spring, and try to depart for the pole as early as possible after that. The provision for support parties for Scott's party was more extensive than Shackleton's had been three years earlier: more men, equipment, supplies, and transport accompanied the polar party for much greater distances on the outward traverse. Successfully navigating Beardmore Glacier, on December 21, twelve men reached Mount Darwin (see Fig. 4.14), where they laid a depot and four of them turned back. On January 4, 1912, three of the remaining support party turned back at 87° 32′ S, and Lieutenant Henry ("Birdie") Bowers joined the pole party of Scott, Wilson, Seaman Edgar Evans, and Captain Lawrence E. G. Oates.

With 168 miles to the pole, the five apprehensively girded for the final stretch and the "appalling possibility" of finding the Norwegian flag flying at the South Pole. On the afternoon of January 16, about twenty miles out from the pole, Scott's party came upon the remains of a camp, a black flag flying from a sledge bearer, and the tracks of men and many dogs. With shattered spirits the men pulled the final distance, arriving at the South Pole on the morning of January 18, 1912. There they found a small tent left by the Norwegians and inside a note with the names of the men who had preceded them to this spot on

December 16: Roald Amundsen, Olav Olavson Bjaaland, Hilmer Hanssen, Sverre Hassel, and Oscar Whisting. Scott's men went through the motions, measured their position with a sun compass, built a cairn, unfurled the "poor slighted Union Jack," and shot a photograph.

In his diary Scott despaired:

> Great God! This is an awful place and terrible enough for us to have laboured to it without the reward of priority. Well, it is something to have got here, and the wind may be our friend to-morrow. We have had a fat Polar hoosh in spite of our chagrin, and feel comfortable inside—added a small stick of chocolate and the queer taste of a cigarette brought by Wilson. Now for the run home and a desperate struggle. I wonder if we can make it.

As history tells us, the party was doomed. With seven hundred miles of the inward march completed, they ground to an agonizing halt eleven miles short of a depot with more than a ton of food, while a howling blizzard held them in their tent for their final days. The party's labor, their triumph and defeat, their ultimate sacrifice, the grace and eloquence of their diaries, the fossil-bearing rocks they carried to the end, and the irony of the final blizzard are the stuff of fame and legend.

Amundsen: Taking the Mountains in Stride

Roald Amundsen is legendary not only for reaching the South Pole first but also for the apparent ease with which he accomplished the feat. When the Norwegians encountered the mountains 250 miles beyond Beardmore Glacier and 100 miles closer to the South Pole, they took them simply as a challenge to be surpassed, not as new lands to be charted with the same care as the British surveyors (Fig. 5.1). Their passage through the mountains was short, but fiendishly steep, and the crevasses on the plateau side were every bit as dangerous as those encountered on Beardmore Glacier by the British. Amundsen's matter-of-fact writing style was the antithesis of Scott's eloquence, belying the rigor and proficiency of his accomplishment, and upon his return some in Britain seized on this nonchalance to denigrate his fame.

The third son of a Norwegian ship owner, Amundsen consciously pursued a career

Figure 5.1. (overleaf) Sweeping for seventy-five miles to the northwest of Mount Griffith, the heart of the Queen Maud Mountains culminates in Mount Fridtjof Nansen, the blocky summit at the center skyline. In 1911 at the foot of that massif, Roald Amundsen's party found passage through the Transantarctic Mountains on its quest for the South Pole. One hundred seventy miles southwest of Shackleton's route on Beardmore Glacier, none could have predicted that the men would discover one of the narrowest connections in all of the Transantarctic Mountains linking the Ross Ice Shelf (at the right skyline) and the polar plateau (at the left). Amundsen Glacier flows from left to right in the middle foreground of the image. This glacier and the surrounding mountains were first seen and mapped from the ice shelf in 1929 by a ground party of Byrd's First Antarctic Expedition (BAE I), led by geologist Laurence Gould.

as "explorer" from the time he dropped out of medical school at the age of twenty-two. After his success through the Northwest Passage, Amundsen promoted his next expedition to the Arctic, where he would sail the *Fram* around Cape Horn and take her into the Arctic Basin through the Bering Strait, there allowing her to be frozen into the ice and then to drift west, emerging ultimately in the North Atlantic. This voyage would be similar to one undertaken in 1893–1896 by Fridtjof Nansen, for whom the *Fram* had been designed and commissioned. The stated purposes were oceanographic and magnetic observations. Although a traverse to the North Pole was not publicly announced, to plant the Norwegian flag at the pole was a strong incentive for support from the donors whom Amundsen approached, and was the single feat that would assure his widening fame.

When the announcements came that the North Pole had been reached, some of the backers withdrew their offers, and new funds stopped coming in. It was the pole, not the science, that most of the patrons cared about. The same was true of Amundsen. In secret he hatched a plan that would ensure his fame in the history of exploration. His only confidants were his brother Leon, Professor Bjørn Helland-Hansen, and Lieutenant Thorvald Nilsen, the commander-designate of the *Fram*. He announced publicly that plans were moving forward for the Arctic expedition, undertook a shakedown cruise in the North Atlantic, and on August 9, 1910, slipped quietly out of the harbor in Christiansand. It had seemed strange that the ship carried ninety-seven dogs when they could have been bought in Siberia at the end of the Pacific voyage, but the pointed questions were ducked well enough to distract those who suspected that all might not be as it seemed.

The last landfall for fresh food and water before the long voyage was Funchal, Madeira. After the ship was loaded, Amundsen called together the eighteen men of his party and offered a challenge, to sail south to Antarctic waters, and from there to vie with Scott and the British in a race to the South Pole. Every man was for it! That night they wrote excited letters to family and friends and gave them to Leon, who would post them once the party was on the high seas and no one could stop the renegades. Leon also handled the press, releasing the story on October 2. Professor Helland-Hansen spoke to Nansen personally about the change in plans. And Amundsen broke the news to Scott with his famous Madeira cablegram.

The *Fram* sailed south from Madeira around the Cape of Good Hope and eastward toward the Ross Sea, aiming to cross latitude 65° S at longitude 175° E. The Norwegians sighted their first iceberg on New Year's Day 1911. By the following evening they encountered the pack just south of the Antarctic Circle and headed into it. After only four days of relatively smooth passage they cleared the pack, and by the afternoon of January 12 lay at the entrance to the Bay of Whales on the eastern edge of the Ross Ice Shelf. From the descriptions of others who had been there, Amundsen guessed that the Barrier was grounded toward a rise to the east of the bay, providing a site for a base that would not on a whim of Nature calve into the Ross Sea.

A site for the base, named Framheim, was chosen on the east side of the Bay of Whales about two and a half miles in from the ice edge on a level névé between large, but old, pressure ridges. They used the dogs to sledge the supplies from the *Fram* to a staging area, began construction of the hut on January 17, and finished it on the 28th, while stores continued to be unloaded. The *Terra Nova* paid its visit on February 4, adding incentive to Amundsen's plan for attacking the pole.

On February 10 the first of three depot-laying traverses began. The party consisted of Amundsen, Kristian Prestrud, Helmer Hanssen, and Hjalmar Johansen. Prestrud was the "forerunner" out ahead of the party, setting the direction and pace while looking for crevasses. The lead sledge was handled by Hanssen, who Amundsen said was "the most efficient dog-driver" he had ever met. Amundsen brought up the rear, where he could watch the rest of the team and catch any objects that might fall from the forward sledges. Each of the three sledges carried about 550 pounds of provisions and was pulled by six dogs.

Although the Barrier was hazy for the first couple of days making surface definition nil, the dogs pulled well. By February 14 the party had reached 80° S. There they deposited more than one thousand pounds of dog food, making a pile of cases twelve feet high and marking it with a bamboo pole and flag. The dogs pushed it hard on the inward journey while the men rode on the nearly empty sledges, making the return ninety-seven miles to Framheim in two long days. Amundsen was justifiably pleased with the dogs' first performance on the Barrier, calling it a "brilliant result," and looked forward with confidence to the ensuing sledge journeys.

With only Henrik Lindstrøm, the cook, left behind, the second depot-laying party left Framheim on February 22, with forty-two dogs pulling seven heavily loaded sledges. On February 27 the party reached the depot at 80° S. This time it was carefully marked by twenty numbered flags with a spacing of three thousand feet, spread on both sides of the depot in an east-west line perpendicular to the direction of the traverse. Amundsen's fear was that the party would drift off course on the return and miss an essential food cache. Unlike the British traverses along the front of the Transantarctic Mountains, where depots had been sited by the bearings or alignment of peaks, on the Barrier there were no such landmarks for determining location.

The party pushed on to 81° S, arriving on March 3. There 1,234 pounds of dog food were left before Olav Bjaaland, Sverre Hassel, and Jørgen Stubberud turned back. The other five continued south to 82° S, where a final 1,366 pounds of provisions, mainly dog food, were cached on March 6. Dropping temperatures during the outward journey coupled with the heavy weights had tested the dogs to their limits. The teams had also experienced their first crevasse field, with three dogs going in together, caught by their harnesses and held by their startled mates. Amundsen had hoped to reach 83° S on this traverse, but the emaciated state of the animals meant that he dared not risk going farther than the depot at 82° S. As it was, eight dogs were lost on the return as temperatures reached below minus 40° F.

The third and final depot party of the season departed on March 31, with seven men, six sledges, and thirty-six dogs. Amundsen stayed at Framheim during this trip. The party, commanded by Prestrud, returned ten days later, having delivered twenty-two hundred pounds of fresh seal meat and sundry other foodstuffs to the 80° S depot, bolstering that cache to forty-two hundred pounds. Combined with the other two depots, the total weight of provisions already in place for the following season was in excess of three tons—nearly three times the cache that Scott had laid at his main depot 150 miles south of Hut Point.

With winter approaching, the nine Norwegians settled into a productive routine at Framheim. Once the last seals had left, the dogs were let loose during the day and then chained before feeding for the night. Separate maternity quarters were constructed for each of the pregnant bitches, which produced a substantial number of pups during the winter. From experiences gained on the three depot-laying traverses, almost all of the equipment was overhauled. Seven of the sledges were reduced in weight from 163 pounds to 53 pounds. The three-man tents that the expedition had started with were modified to five-man tents by cutting out the fronts of two smaller tents and sewing them together. The ski boots were made larger and softer.

For all of his careful planning, one critical item that Amundsen had forgotten to include was a snow shovel. Consequently, a large drift formed beside the main hut without the means of removing it. In due course, Bjaaland fashioned a set of shovels from a supply of steel plates, but the men, instead of digging away the drift, tunneled out a set of rooms from within it, all connected back to the main hut. These included a carpentry shop, a petroleum cellar, a passage to the coal stores, a clothing store, a sewing room, a room for pendulum observations, and the "Crystal Palace," where ski and sledging equipment was stored.

Anxious to set off as early as possible in the spring, Amundsen had the sledges loaded and staged at the starting point on the south side of the Bay of Whales before the sun returned in late August. With the air temperature rising to minus 20° F on September 6 and minus 7° on the 7th, the Norwegians sprang from their base the following day. The expedition included eight men each with a team of twelve dogs, wild to be on the trail—so much so that two of the teams escaped their handlers and ran off before reaching the loaded sledges at the starting point.

The optimistic reading of the temperature was premature. A week later the thermometer had plummeted to minus 68° F. The shivering dogs were unable to keep warm, so Amundsen decided to leave most of the party's load at the 80° S depot and return to Framheim to wait until the season truly turned. The inward march was severe, with three dogs lost and three of the men suffering frostbite in their heels. Each team ran at its own pace, and all were back to the base by September 17.

During the next month, while the frostbite ran its course and a flock of petrels signaled the return of spring, Amundsen made the decision to streamline his polar party to five, and to have the other three men—Prestrud, Stubberud, and Johansen—go exploring in King Edward VII Land.

The polar party (Amundsen, Bjaaland, Hanssen, Hassel, and Oscar Wisting), set off on October 19, 1911, with four sleds, each pulled by a team of thirteen dogs. With the light loads, the men rode on the sledges and covered more than twenty-five miles per day, reaching the 80° S depot in four days. Even with the sledges loaded, the dogs were strong enough to tow the men on skis as well; the men rode behind the entire way to the edge of the Barrier. Between 80° and 82° S the party kept to a strict and easy seventeen-miles-per-day schedule with a mandated two-day rest period at 81° S.

From 80° 23′ S all the way to the pole, the party compulsively built lines of snow beacons across the trail, six feet high, each with a numbered piece of paper indicating position. In all 150 beacons were erected from nine thousand cut blocks of snow. Amundsen defends his "prudence" by saying that one cannot be too careful on featureless ice fields; the project also gave the dogs a rest that kept them running at a fast pace. In fact, the party was making such good time that Amundsen was unconcerned about the time taken up in beacon construction. The system of beacons proved itself worthy soon enough, when the party approached the 82° S depot. In the preceding several days, foggy weather had allowed the party to veer off course, but when a flag on a pole loomed in the haze, the men knew that they were three and a half miles to the west of the depot, and they easily followed the line of flags to it.

On the morning of November 8, the party sighted land, a thin line of summits across their path, "lofty and clear in the morning sun." On the first sledging trip onto the Barrier the season before, the men had joked about the possibility that the Barrier simply went on to the pole, a smooth, rising field of white merging ice shelf and plateau. It was the end run idea that Scott had held to for a while. Shackleton had seen the mountains dissolve out there somewhere around longitude 180° and south of latitude 84°, but one could still imagine the mountains petering out not far beyond that sighting. There had been a lot of speculation on that first run the previous year.

The realistic expectation was that the mountains continued in a southeasterly direction from Shackleton's farthest sighting. If one could project correctly, the mountain front would be more than one hundred miles south of the mouth of the Beardmore Glacier where it crossed 164° W, the meridian Amundsen was tracking. That morning confirmed the expectation that the mountains continued. But continued to where? Would Amundsen's party finally view their termination, or would more mountains materialize from the haze at the southeast horizon?

By the evening of November 8, the party had reached 83° S, where their next depot was laid. Degrees of latitude were slipping by at a rate of one every three days. As they drew closer to the mountains, the features loomed with ever more detail. Straight ahead was a high, blocky massif, freestanding and bold, with bare rock faces faceted across its upper slopes. In honor of his mentor, Amundsen named it Mount Fridtjof Nansen (Fig. 5.2). Immediately to the left (east), beyond a cleft in the skyline, rose another massif perhaps three thousand feet lower in elevation and almost wholly snow clad. He named it Mount Don Pedro Christophersen in honor of a patron living in Buenos Aires who had helped substantially with both provisions and monetary contributions. Farther to the left, more blocky ramparts appeared to be separated by other clefts. Off to the right a long outlet glacier with a gentle meander partway up its length appeared to reach directly to the plateau, but the Norwegians were banking on a shorter crossing through one of the clefts, so they did not deviate from their due-south course.

On November 12 the party reached 84° S, where they laid their next depot. Amundsen records, "On that day we made the interesting discovery of a chain of mountains running to the east; this, as it appeared from the spot where we were, formed a semicircle, where it joined the mountains of South Victoria Land." In other words, the mountains that trended in from the northwest from the Beardmore Glacier area turned in a long arc and ran off to the east.

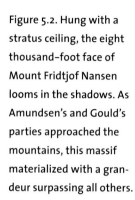

Figure 5.2. Hung with a stratus ceiling, the eight thousand–foot face of Mount Fridtjof Nansen looms in the shadows. As Amundsen's and Gould's parties approached the mountains, this massif materialized with a grandeur surpassing all others.

The following day the party ran twenty-three miles in a thick fog. On November 14 they reached 84° 40′ S. From there the mountains to the east appeared to arc around even more and to trend off to the northeast. The following day the men were building a depot at 85° S, having traversed a series of troughs and swells in the ice, with the mountains almost at hand at the margin of the ice shelf. On November 16 the party crossed two more depressions and swells, both much larger than the previous day, but except for the steepness the going was easy, with only a few old snow-filled crevasses to avoid. At the top of the second swell the men were presented with a smooth snow field rising up into the foothills. They had left the Barrier and were at the doorstep of the mountains, having encountered none of the excess of crevasses at this juncture that others had found farther to the north.

Except for the outlet glacier to the west, named Liv Glacier for Nansen's daughter, the mountains showed no clear gateway to the plateau from this vantage, so the decision was made to do a direct assault, with the hope of reaching the notch between Mount Fridtjof Nansen and Mount Don Pedro Christophersen, seen from farther out on the ice shelf (Fig. 5.3). Here at the fringe of the foothills the men recalculated their weights, allocated a portion constituting sixty days of provisions, and placed the remainder in a depot. They had 683 miles to cover out to the pole and back to this cache. All forty-two of the remaining dogs would be used in the ascent to the plateau, but once there, twenty-four would be shot. By the time the party returned to this depot, the number of dogs would be reduced to twelve, two teams of six, each pulling a single sledge on the final run to Framheim.

After everything was readied, Amundsen and two others put on skis and set out

Figure 5.3. Map of the Axel Heiberg area showing Amundsen's route through the mountains. Campsites are indicated with dots.

for reconnaissance. They headed toward the closest rock, a tiny nunatak about 1¾ miles south of camp that was named Mount Betty, for Amundsen's housekeeper and childhood nursemaid (Fig. 5.4). When they were abreast of it, they split to the west for another three or four miles, ascending over an undulating terrain that headed in small glaciers several thousand feet above at the crest of these foothills. Although it meant a very stiff climb up to a pass, the Norwegians were unwavering in their continued choice of the direct route south.

Their descent back to camp was a gleeful run on skis. When they passed Mount Betty, Amundsen and Bjaaland skied over for a look. Snow reached almost to the top, where they took off their skis and had but a little climbing on bedrock to reach its crest. This was the only landfall of the entire expedition. The rock was dark and rubbly, forming loose scree. After relaxing on the outcrop, the men "photographed each other in 'picturesque attitudes,' [and] took a few stones for those who had not yet set foot on bare earth." But Amundsen relates that these pieces of rock failed to impress the others back at camp. "I could hear such words as, 'Norway—stones—heaps of them,' and I was able to

Figure 5.4. At the front of the photo, detached from its talon, the hooked claw of Mount Betty, Amundsen's only landfall in Antarctica, connects through a gnarly eastern limb back to the massive body of Mount Fridtjof Nansen. Flanked to the south (left) by Axel Heiberg Glacier and to the west (right) by Liv Glacier, the central massif rises abruptly to an elevation of 13,350 feet along its shadowy, northeastern wall. Mount Don Pedro Christophersen is the dark, dome-shaped massif on the far side of Axel Heiberg Glacier. The pair of stepped icefalls between Mount Don Pedro Christophersen and Mount Fridtjof Nansen was the crux of Amundsen's crossing of the Transantarctic Mountains. The route taken the first day in reconnaissance by Amundsen and Bjaaland is illustrated in blue. Amundsen's route south, shown in magenta, links to the route on Fig. 5.5. Gould's route up Strom Glacier to the base of Mount Fridtjof Nansen appears in yellow. The orange lines trace his two ascents of the lower reaches of the massif. Camps are also indicated. The steadily rising track of Byrd's flight to the South Pole is shown in green in the upper reaches of Liv Glacier at the upper right of the image.

Figure 5.5. Amundsen's route through the Transantarctic Mountains winds purposefully across the foothills, up the icefalls of the Axel Heiberg Glacier, and behind Mount Engelstad, the low pyramid to the left of Mount Fridtjof Nansen. The first night's camp "lay on a little glacier among huge crevasses." The blue lines show the reconnaissance routes of Wisting and Hanssen to the right and Bjaaland to the left, with both parties reporting back that the next day they would have to descend. The steepest bit of climbing of the entire traverse was in the shadowed stretch of the ridge in the middle of the image. The Norwegians' three camps at the base, middle, and top of the icefalls are indicated. The topmost in the gap to the left of Mount Engelstad was the "Butcher's Shop." Mount Balchen, occupied by the Topo East survey during Operation Deep Freeze 63, is the prominent shoulder immediately in front of the summit plateau of Mount Fridtjof Nansen.

piece together and understand what was meant. The 'presents' were put in depot, as not absolutely indispensable on the southern journey."

On the morning of November 17, the party made its first volley at the mountains, crossing the undulating front of the small range and then climbing partway up its side. At the end of the day the men had driven eleven and a half miles and risen two thousand feet. Their camp that night was about halfway up the mountain and lay between several sizable crevasse fields (Fig. 5.5). After dinner Wisting and Hanssen went in one direction and Bjaaland went in another to climb to the crest and scout the route for the following day. When they arrived back at camp, they all agreed that once past the ridgeline the only way to go was down, but each claimed to have found the better route. After listening to both sides, Amundsen chose the pass through the lowest part of the ridge directly at the head of the glacier where they camped.

The next day the dogs performed magnificently, pulling their sledges up the steep slope without the need to double-team, even though the snow was deep and soft. At the pass Amundsen beheld what the three others had viewed the night before. "We now saw the southern side of the immense Mt. Nansen; Don Pedro Christophersen we could see in his full length. Between these two mountains we could follow the course of the glacier that rose in terraces along their sides. It looked fearfully broken and disturbed, but we could follow a little connected line among the many crevasses; we saw that we could go a long way."

The glacier was named the Axel Heiberg, for a wealthy patron of many Norwegian expeditions. From the pass a steep descent of about eight hundred feet down the valley wall put the party on a steep glacier that flowed down to the left and merged with Axel Heiberg Glacier. The ridge that confined the tributary glacier on its far side, however, obscured the view of the connection between the upper and lower portions of Axel Heiberg Glacier, and Amundsen decided that rather than descend, they would cross the far ridge setting a straight course toward the terraced icefalls to the south of Mount Fridtjof Nansen.

The ascent out of the valley was "the steepest bit of climbing on the whole journey—stiff work even for double teams." Having climbed 1,250 feet and reached an elevation of 4,550 feet, the party finally could see what lay between them and the icefalls, a broad outlet glacier with deep longitudinal furrows. The swaths between the furrows showed signs of many crevasses, but the furrows appeared to be filled with snow. In particular, one line near the middle of the glacier seemed to connect right up to the foot of the first icefalls. Here in one vista was the entire width of the Transantarctic Mountains, foothills rising through ridge systems up to blocky massifs that bounded the plateau (see Fig. 5.5). The only other place that the mountains were so narrow was around David Glacier and the Drygalski Ice Tongue. Here one could see the flat horizon of the plateau between Mounts Nansen and Christophersen, twenty-five miles away at most. The Beardmore had been one hundred miles of gently rising glacier riddled with crevasses. The Axel Heiberg would be a short glacier approach, hopefully without many crevasses, followed by a searing set of icefalls that would challenge for only a few days. But during those few days the grit, the mountaineering skills, and the luck of the Norwegians would be tested beyond any challenge yet posed by the continent.

Braking their sledge runners with rounds of rope, they descended to Axel Heiberg Glacier, piecing together a route that took them over a soft snow surface to the white furrow and thence to the foot of the first icefall. There they camped close to the base of Mount Don Pedro Christophersen, across from the towering, snow-covered, south wall of Mount Fridtjof Nansen, rising nine thousand feet out of the glacier directly to the north. That evening Hanssen and Bjaaland did a reconnaissance of the icefall, reporting that the first terrace would be accessible the following day.

As always on the more demanding ascents, Bjaaland was the forerunner, scouting the trail and leading the climb. The sledges were double-teamed up the steep icefall, but because the surface was harder, the dogs were able to grip better than in the deep snow. Although crevasses were everywhere, a connected trail was cobbled out of safe ground and solid bridges. Before midday the party had reached the first terrace. The second icefall loomed ahead with "nothing but crevasse after crevasse, so huge and ugly" that this step was impassable. However, to the left the terrace seemed to rise gently toward Mount Don Pedro Christophersen and to merge into its snowy lower slope. The men headed in that direction but soon found themselves in a cul-de-sac of open chasms, so they camped.

While the dogs were being fed and bedded and the tent set up, Amundsen led a reconnaissance sortie to the base of an ice ridge above the camp. The crux of the traverse was a crevasse bridge no wider than a sledge that crossed a chasm of deep blue opening on both sides. The men returned to camp for dinner and then went out again to see what lay beyond the ridge. By keeping in close under Mount Don Pedro Christophersen, they found the passage above the ice ridge to be smooth, except for a few large, open crevasses that were easily avoided. Before long they were sure that they had passed the chaotic part of the glacier and that only one final ice rise stood between them and the plateau. The best route over the rise was uncertain, so the men pushed on across the terrace, scouting the passes between the large beehive-shaped summits at the head of the Axel Heiberg catchment. The least steep ascent appeared to be to the north between Mount Fridtjof Nansen and Mount Ole Engelstad.

With confidence that they would be on the plateau the following day, the three men skied back to camp. As they came out on a rise and looked down at their tent, Amundsen reflected on his being in this daunting reach:

> The wildness of the landscape seen from this point is not to be described; chasm after chasm, crevasse after crevasse, with great blocks of ice scattered promiscuously about, gave the impression that here Nature was too powerful for us. Here no progress was to be thought of.
>
> It was not without a certain satisfaction that we stood there and contemplated the scene. The little dark speck down there—our tent—in the midst of this chaos, gave us a feeling of strength and power. We knew in our hearts that the ground would have to be ugly indeed if we were not to manoeuvre our way across it and find a place for that little home of ours.

During the night avalanches rumbled down from the upper slopes of both Mount Fridtjof Nansen and Mount Don Pedro Christophersen. On the morning of November 20, the weather was still and clear. The pull up to the next terrace was strenuous, but the dogs managed to do it with single teams. Once up, the party aimed straight for the northern foot of Mount Ole Engelstad. As they rounded it, the plateau opened before them beyond a final steep rise. In a switch from the rigid lunch routine of dry biscuits, Hanssen broke out the primus and cooked up a pot of thin chocolate, while the men contemplated what lay ahead. As if they knew this would be their last steep pull, the dogs dug in after the lunch break and reached what appeared to be the plateau on the south side of Mount Ole Engelstad. It was now time to turn south again. Directly ahead was a snowy ridge that projected to the west from Mount Don Pedro Christophersen. As they ran up onto it, the surface changed from the soft snow that had been with them since they reached the mountains to hard, sharp-edged sastrugi trending northwest-southeast. When the gratified party camped that night at 8:00 P.M., it was at 10,920 feet, having covered 19¼ miles and risen 5,750 feet.

Figure 5.6. (opposite) Shaded-relief map of the Queen Maud Mountains showing the routes taken by the first parties to the area. The dashed insets indicate the locations of Figs. 5.3 and 6.2.

The dogs had performed spectacularly, and now twenty-four would be put to their deaths. The ones to be killed and the ones to be spared had all been decided beforehand. While Amundsen tried to distract himself by preparing the evening meal, each of the drivers shot the chosen dogs from his team, dressed them out, and fed the entrails to the voracious survivors. Although they had been looking forward to the addition of canine cutlets to their diets, that night none of the men had an appetite for dog, in the camp they called the "Butcher's Shop."

The plan had been to spend two nights at this campsite, rearranging provisions onto three sledges, and preparing the dog meat. By the second day the men were ready for their cut of the flesh. Chef Wisting prepared the soup du jour, a mongrel hash with vegetables picked from pemmican, and an entrée of fried cutlets that had the men singing for more. Amundsen ate five. During this rest the surviving dogs also ate themselves full. Stormy weather caused delay, so by November 25, when the party finally left for the pole, everyone's hunger had been satisfied.

Sitting out the storm that raged unabated, the men became increasingly restless. By the fifth day, they all agreed that driving in the storm would be better than being pinned down, so they marked their depot with Hassel's sledge and a broken ski, and off they went into the teeth of the wind. Amundsen revealed himself as he reminisced on this episode. "When I think of my four friends of the southern journey, it is the memory of that morning that comes first to my mind. All the qualities that I admire most in a man were clearly shown at that juncture: courage and dauntlessness, without boasting or big words. Amid joking and chaff, everything was packed, and then—out into the blizzard."

The weather for the next two weeks was never good. Occasional clearings would briefly open, but typically the party was engulfed in either fog or blowing snow. The day that they left the Butcher's Shop, they put in 11¾ miles and dropped 620 feet. The following day they reached 86° S, 18½ miles farther and an additional 825 feet of descent. Something wasn't right. The plateau was supposed to rise toward the pole, but here it was dropping off. Two days later, November 28, the party hit crevasses and the men decided to make their next depot, now 140 miles south of the Butcher's Shop. Intent on their true-south heading, the men pushed into the crevasse field that seemed never to end. With visibility nil most of the time, the clouds would occasionally part, revealing the danger that surrounded them. "Devil's Glacier" was the party's unanimous choice in naming this passage (Fig. 5.6).

On December 1 the ice began to rise, and by the 3rd it seemed that the party had finally cleared the crevasses, as well as the foul weather, but one last batch of crevasses forced them to stop that afternoon, while Amundsen and Hanssen scouted a route into the "Devil's Ballroom." Negotiating this last obstacle was taxing, with open chasms riddling the field, spawning a number of near catastrophes. In describing the crux where a narrow bridge crossed a deep divide, Amundsen departs from his usual understatement: "The crossing of this place reminded me of the tight-rope walker going over Niagara."

This was the final impediment the party faced. By the end of the day they had truly reached the plateau, with smooth sailing to the South Pole. During the occasional clearings the previous week, stark mountains had revealed themselves through the mists. On November 27, when the fog briefly lifted, a pair of long, narrow, snow-clad ridges had

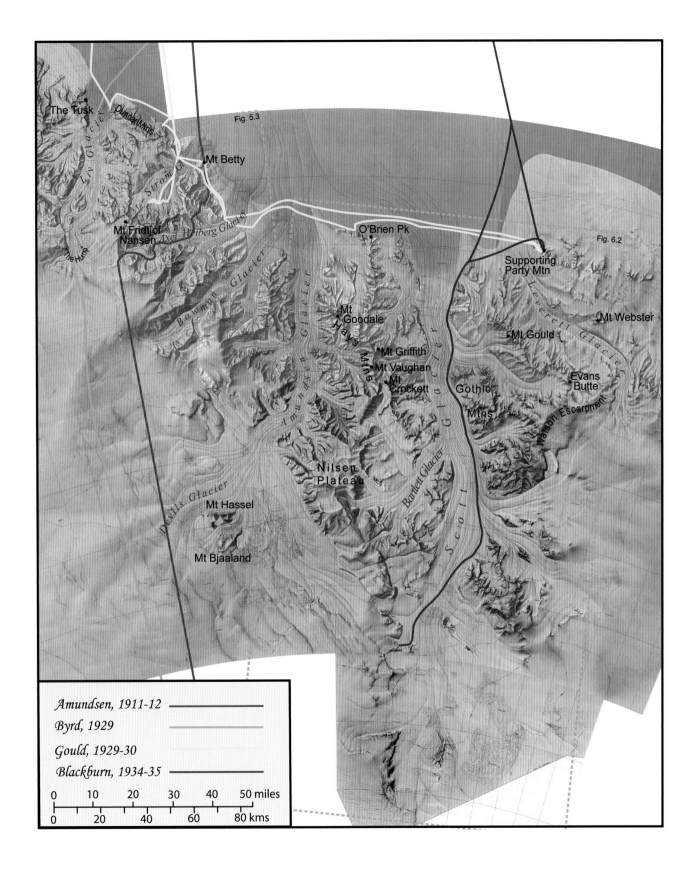

The Tusk
Liv Glacier
Duncan Mtns
Strom Gl
Mt Betty
Fig. 5.3
Mt Fridtjof
Nansen
Axel Heiberg Glacier
The Hump
O'Brien Pk
Fig. 6.2
Supporting
Party Mtn
Mt Webster
Bowman
Glacier
Mt
Goodale
Mt Gould
Leverett Glacier
Amundsen Glacier
Hays Mtns
Mt Griffith
Mt Vaughan
Mt
Crockett
Gothic
Mtns
Evans
Butte
Watson Escarpment
Nilsen
Plateau
Bartlett Glacier
Scott
Devils Glacier
Mt Hassel
Mt Bjaaland
Glacier

Amundsen, 1911-12
Byrd, 1929
Gould, 1929-30
Blackburn, 1934-35

0   10   20   30   40   50 miles
0   20   40   60   80 kms

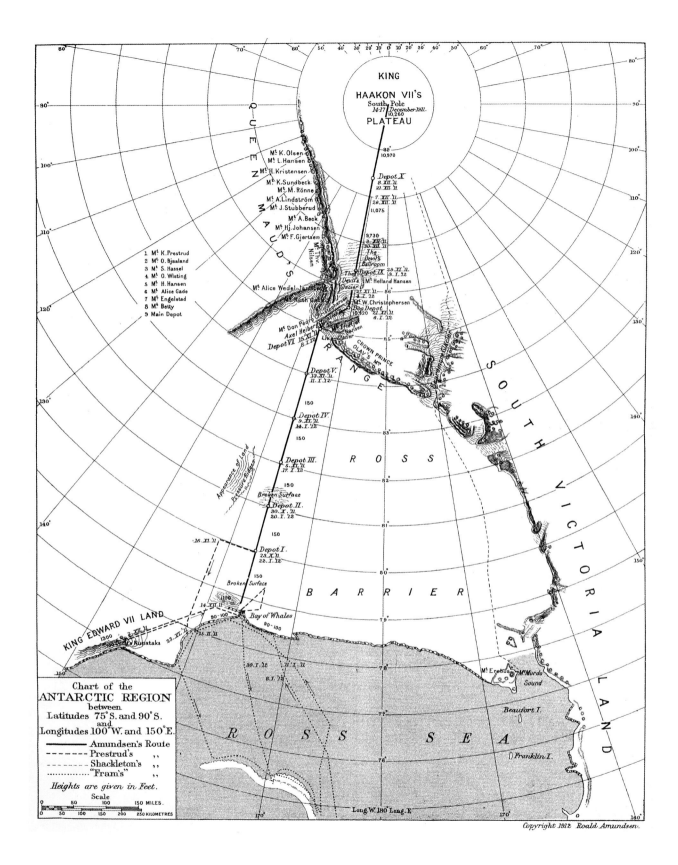

KING
HAAKON VII'S
South Pole
14–17 December 1911.
10,260
PLATEAU

Q U E E N   M A U D ' S

Mt K.Olsen
Mt L.Hansen
Mt H.Kristensen
Mt K.Sundbeck
Mt M.Rönne
Mt A.Lindström
Mt J.Stubberud
Mt A.Beck
Mt Hj.Johansen
Mt F.Gjertsen

1 Mt K.Prestrud
2 Mt O.Bjaaland
3 Mt S.Hassel
4 Mt O.Wisting
5 Mt H.Hansen
6 Mt Alice Gade
7 Mt Engelstad
8 Mt Betty
9 Main Depot

Mt Alice Wedel Jarlsberg

Mt Don Pedro Christophersen
Axel Heiberg
Depot VI. 15.XI.11
8.I.12

Depot X
8.XII.11
23.XII.11

7.XII.11
26.XII.11
11,075

9,730
3.XII.11
30.XII.11
The
Devil's
Ballroom
The Depot IX
Devil's   3.I.12
Glacier II   Mt Helland Hansen
27.XI.11   86
1.I.12
Mt W.Christophersen
Dog Depot
10,920
6.I.12

R A N G E

CROWN
PRINCE
OLAV'S Mts

Depot V.
13.XI.11
11.I.12

150

Depot IV.
9.XI.11
14.I.12

150

Depot III.
5.XI.11
17.I.12

150

Broken Surface

Depot II.
30.X.11
20.I.12

150

Depot I.
23.X.11
22.I.12

150

Broken Surface

14.XII.11   1100
90–100

Bay of Whales
90–100

KING EDWARD VII LAND
1300
Nunataks
25.XI.11
15.II.11

90.I.12   11.I.11
8.I.12

R O S S

S O U T H   V I C T O R I A   L A N D

B A R R I E R

79°

78°

Mt Erebus
Mt Murdo
Sound

77°

Beaufort I.

76°

Franklin I.

R O S S   S E A

Long.W. 180 Long.E.

170°

170°

160°

Chart of the
ANTARCTIC REGION
between
Latitudes 75° S. and 90° S.
and
Longitudes 100° W. and 150° E.

———— Amundsen's Route
– – – – Prestrud's   ,,
– - – - Shackleton's   ,,
·········· "Fram's"   ,,

Heights are given in Feet.

Scale
0   50   100   150 MILES.
0   50   100   150   200   250 KILOMETRES

loomed to the west, and then they were gone. These were named the Helland Hansen Mountains and noted as the only mountains to the party's right once they had reached the plateau (Fig. 5.7). On the 28th new mountains had appeared intermittently to the southeast. The following day the weather cleared, offering better views. The closest land was a set of four rounded massifs, which stood out separately from the mountains behind them. Amundsen named the quartet after his steadfast sledgemates, Hanssen, Wisting, Bjaaland, and Hassel, although today only the names Mount Hassel and Mount Bjaaland remain on the pair of islands to the southeast of Devil's Glacier.

The mountains beyond were a magnificent line of peaks that trailed off into cloud at their southeastern extremity. Amundsen estimated them to be ten thousand to fifteen thousand feet high (we know today that the lower estimate is their highest) and recorded his impressions: "Peaks of the most varied forms rose high in the air, partly covered with driving clouds. Some were sharp, but most were long and rounded. Here and there one saw bright, shining glaciers plunging wildly down the steep sides, and merging into the underlying ground in fearful confusion."

The last appearance of the mountains was the day the party left the Devil's Ballroom, reaching out to a vanishing point at 87° S. On December 4–6, the party covered twenty-five miles each day, the latter two days under full blizzard conditions. With a shout on the morning of the 7th the party hoisted the Norwegian flag onto Hanssen's sledge as they passed Shackleton's farthest south, 88° 23′ S. Day after day they drew closer to their goal, wondering whether they would find footprints of Scott's party welcoming their arrival.

On December 14, when the men awakened, as always, at 6:00 A.M., they found a fine, clear morning. Perhaps with an extra swagger to their gait, the men knew that this would be the day. By 10:00 A.M., the sky had clouded over. When a noon reading of the sun was not possible, they pushed on by dead reckoning. Amundsen recorded the moment:

> At three in the afternoon a simultaneous "Halt!" rang out from the drivers. They had carefully examined their sledge-meters, and they all showed the full distance—our Pole by dead reckoning. The goal was reached, the journey ended.

Ceremoniously, with all five hands on the staff, the Norwegians planted the flag in the name of King Haakon VII.

The men made camp and retired for a spartan celebratory dinner, with a small piece of seal meat each. After dinner Amundsen pulled out a pipe that carried the names of many Arctic landmarks. On this he carved "South Pole," and Wisting pulled out a secret plug of tobacco that he gave to the leader as a gift. Everyone turned in for a few hours of sleep, then was back outside again for a midnight sighting on the sun. Their calculations put the latitude at 89° 56′ S.

Amundsen's plan from here was methodical. They would "encircle" the South Pole by sending three men out 12½ miles on paths at right angles to each other and the inward route. While Wisting, Hassel, and Bjaaland marched out their lines, Amundsen and Hanssen rearranged the provisions onto two sledges and undertook a series of hourly sun sightings to accurately determine their longitude.

On the morning of December 16, the lightened party moved toward the recalcu-

Figure 5.7. (opposite) Amundsen's map of new lands and his track to the South Pole. Refer to the text for details of the geography.

145

lated pole. At 11:00 A.M., they reached the still turning, the spot where all the meridians converge and every direction is north. To record the validity of their location they took hourly sun sightings for a full twenty-four hours. A small tent, which they called Polheim, was securely erected, items were left inside, the flag was hoisted, heads were bared, and a photograph recorded the historic occasion.

One can say that Amundsen and his men lingered at their endpoint, enjoying their rest and their reflection. This was in marked contrast to the British expeditions that reached toward the poles. David: "a right-about turn, and as quick a march as tired limbs would allow back." Shackleton: "We stayed only a few minutes, and then, taking the Queen's flag and eating our scanty meal as we went, we hurried back." Scott (one month hence): "Well, we have turned our back now on the goal of our ambition and must face our 800 miles of solid dragging—and good-bye to most of the day-dreams!"

When the party did finally commence the inward journey, the going was "splendid and all were in good spirits." The weather was much clearer, so the crevassed areas were more easily gotten through, and the dogs pulled as if they knew they were going home. When the mountains that struck off to the southeast began to reappear, Amundsen was curiously unable to recognize the summits that they had measured on the way south. In addition, he now sighted the termination of the mountains at 88° S where they faded into the horizon.

The beacons that had been so obsessively constructed on the outward journey were picked off systematically on the return. The descent of the Axel Heiberg icefalls was accomplished in a day. From the bottom, the party went around the end of the range that they had taken with a direct assault the previous month, and sledged up to their depot at 85° S. After they had finished reloading the sledges, Amundsen and one other man skied up to Mount Betty, where they built a cairn, left seventeen liters of paraffin, some matches, and a letter recounting what they had accomplished. Amundsen also gathered some more rock samples, the only scientific specimens collected during the entire expedition. In all "about twenty" samples were returned, including vein quartz, metamorphic rock, and several varieties of granite.

On January 7, with a full blizzard blowing but with the wind at their backs, the five men and their eleven remaining dogs hauled onto the Barrier with Framheim directly north at the end of their route. Weather was mixed, good days and bad days, but the teams managed between seventeen and thirty-four miles per day, gathering more food than they could eat as they went along. January 18, a fine clear day, was one of note. The party was abreast of a set of high-pressure ridges off a number of miles to their right (west), which they had sighted less clearly on the outward journey.

Unexpectedly, land appeared beyond these ragged ripples in the ice. Amundsen wrote in his diary,

> Great was our surprise when, a short time after, we made out high, bare land in the same direction, and not long after that two lofty, white summits to the south-east, probably about 82° S. It could be seen by the look of the sky that the land extended from north-east to south-west. This must be the same land that we saw lose itself in the horizon in about 84° S., when we stood at a height of about 4,000 feet and

looked out over the Barrier, during our ascent. We now have sufficient indications to enable us without hesitation to draw this land as continuous—Carmen Land.

Upon later consideration, Amundsen waffled on the reality of his sighting of land on January 18. He reaffirmed his earlier observations of the mountain front arcing to the northeast between 86° and 84° S, but restricted the name Carmen Land to that segment. On Amundsen's published map, the sighting of the 18th is recorded as "Appearance of Land." Carmen Land is shown as a straight line of mountains trending northeast from Mount Alice Wedel-Jarlsberg, although for some reason the name was not included on this map (see Fig. 5.7). Subsequent maps of the period, however, did chart Carmen Land (Fig. 5.8).

The party made it back to Framheim at 4:00 A.M. on January 25, having been gone ninety-nine days and covered 1,860 miles. Creeping into the bunk room, they startled their sleeping mates, who rose with warm congratulations, put on the coffee pot, and whipped up a heaping stack of hotcakes. *Fram* had arrived on January 8, but weather had forced her out into the Ross Sea. The day after the polar party returned, she sailed back into the Bay of Whales, bringing the best possible news that the king had supported Amundsen's change of plans and Fridtjof Nansen had concurred. Five days were all it took to complete the packing of the *Fram,* and the expedition set sail for Hobart.

Although Amundsen returned to a hero's welcome in Norway, once the word of Scott's tragedy reached the outside world, Amundsen's success was vilified by many in Britain. He somehow had broken the rules of fair play. He had heartlessly killed his dogs. He had been extremely lucky to find such a narrow passage through the mountains. He had not undertaken any scientific investigations. He was less noble.

Of course, the rules of fair play when it came to planting the flag had always accorded propriety to the one who gets there first. The British had been very good at planting flags for centuries. The superiority of dogs for polar traversing was proven resoundingly, and the key to the success was that the teams were driven by men of the north country, who had been on skis since the time they could walk, and had years of experience handling dogs. That the dogs were killed was calculated and impassive, but Amundsen's soft side can be seen in his assigning the killing to the others in the party, and in his book *South Pole,* when on numerous occasions he speaks of his admiration for the dogs.

To say that Amundsen was lucky to find the Axel Heiberg Glacier is true, but this undercuts what is perhaps the most spectacular accomplishment in the annals of Antarctic mountaineering. Finding the route through the Axel Heiberg icefalls and maneuvering the dog teams through the maze were bold and required the utmost skill.

If one wants to argue that the lack of scientific investigation was a shortcoming of the expedition, it may also be said that science was not part of the plan. The original proposal of sailing the *Fram* into the Arctic Sea was for the purpose of oceanographic investigation, and the *Fram* did in fact conduct oceanographic measurements on its cruise back to Buenos Aires in 1912. But the twenty-odd rock specimens that were returned were a paltry sampling of the new lands the Norwegians had discovered.

Moreover, the map that Amundsen produced was a sorry representation of the country he had surveyed (see Fig. 5.7). To be sure, the discovery that the Transantarctic Moun-

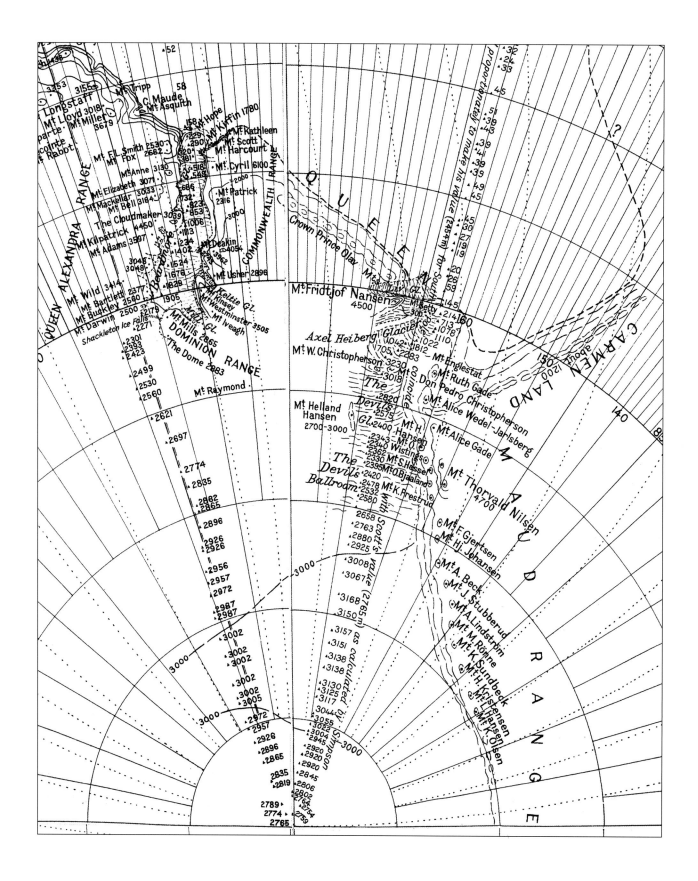

tains extended at least to longitude 150° W advanced the discoveries of Shackleton another 250 miles into the unknown. Although the better part of the distance between Beardmore and Liv Glaciers had not been seen, Amundsen's connecting the mountain front between these two areas was reasonable speculation, accounting for a span of mountains more than 1,000 miles long. The surveying of the line of peaks that Amundsen showed extending southeast from Mount Thorvald Nilsen was so poorly constrained, however, that, when the modern maps were made from aerial photography, the locations of many of the peaks could only be guessed at, and the assignment of names in this area became arbitrary. The northern portion of these mountains is now called the Nilsen Plateau, which peters out to the south in a row of subdued peaks at the head of Scott Glacier. It is possible that some of these distant peaks were visible on the party's return across the plateau, but they extend barely south of 87° S, rather than to 88° S, as Amundsen mapped. The nunataks named for the polar party are not shown on his map, although they are on contemporary maps. Likewise, Mount Ole Engelstad and Mount Wilhelm Christophersen, two summits in the headreaches of Axel Heiberg Glacier, are mentioned in Amundsen's text but not placed on his map. Modern mapping has also shown that Mount Helland Hansen does not exist. A broad undulation of the descending ice cap to the west of his route may be what was mistaken for mountains. As a guide to future explorers in the region, Amundsen's map was seriously wanting.

Whither Carmen Land? Byrd's Flight and Gould's Traverse

Not since the United States Exploring Expedition of 1838–1840 commanded by Lieutenant Charles Wilkes had the Americans ventured into the Antarctic arena. But after World War I and its emergence from isolationist policies, the country was flush with optimism and a new self-image as international player. Although the South Pole had been reached, large spaces on the Antarctic map remained blank and unclaimed, so it reasonably followed that public interest turned to the "Great White South" as a place to demonstrate American prowess. Great advances had been made during the war in aviation and in radio communication, technology that would launch the next era of Antarctic exploration.

From this period emerged a man who for Americans would be synonymous with Antarctica, as were Scott and Shackleton for the British, Mawson for the Australians, and Amundsen for the Norwegians. This man was Richard E. Byrd, a naval aviator during World War I, who had pioneered in navigational techniques beyond the sight of land. On May 9, 1926, he navigated the first flight to the North Pole. The following year he succeeded in a transatlantic flight from New York to Paris. Then, in early 1928, Byrd announced plans to lead an expedition to Antarctica, spearheaded by airborne exploration of uncharted lands and a flight to the South Pole.

In a nine-month frenzy of promotion and organization, Byrd's expedition took form. All donations came from the private sector, the principal backers being Edsel Ford and John D. Rockefeller. The flagship of the expedition was a reconditioned Norwegian whaler recommended to Byrd by Amundsen. Rechristened the *City of New York*, it was supported by a steel cargo ship renamed the *Eleanor Bolling*. Two whaling vessels,

Figure 5.8. (opposite) In this detail of the continental map of Antarctica, published in 1928 by the Geographical Society of New York, Carmen Land is out of kilter with the other names on the map, and the elevation points on Amundsen's traverse are corrected using the elevation for the South Pole determined by Scott. Reproduced by permission of the American Geographical Society.

the *James Clark Ross* and the *C. A. Larsen,* also carried some of the personnel, dogs, and equipment to the Antarctic. The pride of the expedition was three airplanes, including a Fokker Universal, a Fairchild with folding wings, and a Ford trimotor named the *Floyd Bennett,* in honor of the North Polar pilot who had died of pneumonia before the end of that expedition.

The *New York* and the *Bolling* left Dunedin on December 2, 1928, reaching the pack on December 10. There in rough seas ninety tons of coal were transferred in sacks to *New York* before *Bolling* turned back. Next the whaler *Larsen* rendezvoused with the *New York* and towed her through the rough pack. On December 23 the ships broke into open water, *Larsen* dropped her line, and *City of New York* sailed south. The expedition put in at the Bay of Whales on December 28. The site for the base, which was named Little America, was chosen about eight miles onto the ice shelf from where *New York* moored and about four miles north of the Framheim site, though no evidence of the Norwegians' presence was found. A reloaded *Eleanor Bolling* arrived at the Bay of Whales on January 27, was unloaded in five and a half days, and promptly returned to Dunedin.

When *New York* departed for New Zealand on February 28, forty-two men remained behind, as had others before them, to face the winter night. The aerial campaign was two-pronged. One was to the north and east, exploring the unknown coastal area between King Edward VII Land and the Antarctic Peninsula. The other was to the South Pole, basically retracing Amundsen's route across the ice shelf, through the mountains, and onto the polar plateau. A ground party would use dog teams to reach the mountains, where they would radio weather reports for the polar flight, and once it had been completed, would travel east along the mountain front surveying, mapping geologically, and exploring Carmen Land.

In choosing dogs for the overland traverses, Byrd had taken a lesson from the Norwegians. The primary dog handlers for the expedition were three classmates from Harvard: Eddie Goodale, Freddie Crockett, and the ringleader, Norman Vaughan. Vaughan had learned to work with dogs during a nine-month period assisting Sir Wilfred Grenfell, a British physician who had left England in 1892 to serve the medical needs of the Eskimo of Newfoundland and Labrador. The "Three Musketeers," as the Harvard boys were called, trained the year before the expedition in New Hampshire under the guidance of Arthur Walden, an innkeeper and breeder of sled dogs, whom Byrd had enlisted to provide the canine component of the logistics. Although the airplane proved during this expedition that it could reach great distances and record new terrain through aerial photography, the most reliable ground travel remained by dogs. A Ford snowmobile was used to haul depot supplies onto the ice shelf, but it broke down and was abandoned on its maiden run seventy-five miles south of Little America.

Throughout the base-building phase, the dog teams ran a four-mile shuttle from the edge of fast bay ice, where *New York* and *Bolling* had offloaded, to Little America, where the base was taking shape. During this period a number of the men tried a hand at mushing the dogs. One depot-laying party ventured south in early March, as much to give the men and dogs the experience of traversing as to lay out supplies. During a weeklong period they weathered a pounding blizzard, dropped depots at twenty-, forty-, and forty-four-mile distances, and were feeling in such good condition that they did the run back to Little America in one long day.

The same week a sledging incident occurred closer to home. While making a run back from the cache left by the ships, a young surveyor named Quin Blackburn was caught in a sudden squall that brought forty-mile-per-hour winds and thick, blowing snow. Another team, hurrying back along the flagged route to Little America, had seen him disappearing into the drift unable to hold his dogs on course. They also reported that he had not been carrying a sleeping bag, which added to the alarm as time passed and he didn't appear. A search was mounted, which as the hours dragged on came to involve nearly everyone at the base. When Norman Vaughan's group finally found him he was in the lee of his sled in a hole dug in the snow, snugly surrounded by huskies and covered by gasoline cans. Blackburn had been out for eight hours but was none the worse for wear. Byrd duly noted his poise and would enlist this surveyor on his second Antarctic expedition, when he would have him lead a traverse into the very heart of the Transantarctic Mountains.

Like the British before them, Byrd's expedition held high the banner of science. Included were a physicist, a meteorologist, a biologist, and a geologist, as well as several surveyors and topographers. The geologist, second in command of the expedition, was Laurence M. Gould, then assistant professor of geology at the University of Michigan, who had come to polar exploration through a University of Michigan expedition to the Greenland ice cap in 1926 and the Putnam Expedition to Baffin Island in 1927. Gould's principal undertaking of the expedition was to be an investigation of the geology of the Queen Maud Mountains and Carmen Land beyond, as leader of the ground party in support of the polar flight.

During the spring the geological party and a supporting party (led by Walden) each completed two runs, laying a series of depots to 81° 45′ S. The final departure of the geological party was on November 4, 1929. In addition to Gould, who was responsible for navigating, cooking, and geology, the party included Edward E. Goodale; George A. Thorne, topographer; J. S. O'Brien, surveyor; Fredrick E. Crockett, radio operator; and Norman D. Vaughan, chief dog handler. Each of the men except Gould handled a team of dogs. The first two weeks were easy going, with reliance on the well-stocked line of depots. On November 18 the party was buzzed by the *Floyd Bennett,* on its way to laying a depot at the foot of the mountains. As the trimotor passed, Byrd dropped a packet of messages attached to a small parachute. While flying between 81° and 82° S, Byrd searched with binoculars to the west in vain for the "Appearance of Land" with its two snowy peaks that Amundsen had mapped (see Fig. 5.7) and included in Carmen Land in his narrative.

As the plane continued south a line of faint mountains appeared to the southwest, more than 150 miles away, "bending in a broad sweeping curve to the east" and eventually crossing the line of flight. Byrd recorded, "Slowly, now, the Queen Maud Range came into view: first a few lone peaks dancing above the cylinder heads in the arc of the propeller; than dark shoulders of rock draped with snow; then, finally, a solid mass of mountains cut and riven by glacial streams [Fig. 5.9]. Here, indeed, is what we had come so far to see."

In a great panorama of increasing detail the mountains materialized before the advancing plane, but for much longer than Byrd expected none of the landmarks that

Figure 5.9. Liv Glacier spills down from the polar plateau, winding for forty-five miles through buttressed ridges of the Queen Maud Mountains before entering the ice shelf. Byrd's flight up the glacier favored the left (east) side, keeping to the left of the three prominent nunataks in midstream in the upper reaches of the glacier (compare Fig. 5.6). At the termination of the long, dark ridge extending out to the right of Mount Fridtjof Nansen, the glacier rises abruptly at an icefall. This was The Hump, where the passengers feverishly scuttled any loose weight, including survival gear, and prevailed in gaining the altitude necessary for mounting the plateau. Following the flight, Gould's party moved into the end of the dark ridge in the left foreground. From there the men attempted to traverse up Liv Glacier but were thwarted by severe crevassing, after which they turned east (left) and carried along the mountain front.

Amundsen had described were recognizable. Then he spotted Mount Ruth Gade from one of Amundsen's photos that he was carrying, and the rest of the landscape fell into place: Mount Don Pedro Christophersen, followed by Mount Fridtjof Nansen, with the Axel Heiberg cascading between them, Liv Glacier off to the right of Fridtjof Nansen, and beyond that a "consolidated mass, with the frowning ramparts of a fortress," more than twenty miles long rising behind the Prince Olav Mountains. Ashley McKinley recorded the entire passing spectacle with aerial photography.

The plan had been to land near the mouth of Axel Heiberg Glacier, where a depot would be laid for the geological party; the glacier surface, however, appeared too rough in the reflected sun, so the plane flew to the mouth of Liv Glacier, where a relatively smooth snow surface was picked for a touchdown. Dean Smith eased the *Floyd Bennett* onto the sharp-edged sastrugi. While he kept the engines turning, Byrd, McKinley, and Harold June built a snow cairn and laid the depot. Then the plane took off and headed east along the front of the Queen Maud Mountains, McKinley photographing the face of the mountains as they went. Byrd's intention was to fly at least one hundred miles to the east to investigate Carmen Land; but no sooner had they started than June realized that the fuel tanks were alarmingly low. The chewing gum and tape that had been used at the beginning of the trip to stop a leak in one of the hand fuel pumps apparently had not held, forcing an immediate turn toward Little America. It was not until three days later that the geological party learned from the radio that the plane had run empty eighty miles short of Little America, where it landed safely and subsequently was refueled by one of the other planes.

The geological party pushed forward under marginal weather. On November 26 land appeared to the southwest, and by evening the men had spotted Mount Fridtjof Nansen to the south. As the party continued, it steered toward a large gap in the mountains, which at first they took to be Axel Heiberg Glacier but which later proved to be the Liv.

The polar flight was delayed for more than a week by poor weather, either at Little America or on the ice shelf. Finally, on November 28, Goodale radioed Little America that weather was fair and clear. The pole party took off in the *Floyd Bennett* that afternoon with the men in their places: Bernt Balchen, pilot; Harold June, copilot; Ashley McKinley, mechanic and aerial photographer; and Byrd, navigator. The geological party made camp and monitored the radio in case of emergency. At 8:15 P.M. the plane spotted the ground party, "a cluster of little beetles," and dropped them another packet of messages and letters, along with some film and aerial photographs of the mountains shot by McKinley on the depot-laying flight.

They completed the polar flight, though not without travail, for as the crew flew up Liv Glacier (see Figs. 5.4, 5.9), its chosen route through the mountains, it fought a strong downdraft from the plateau and a ground surface rising faster than the plane. Only after scuttling 250 pounds of survival food could the plane gain the altitude to clear the rising surface with little room to spare.

Throughout the passage of Liv Glacier, McKinley recorded the mountains and glaciers out the left (east) side of the plane with still photography and motion pictures. In the stretch passing Mount Fridtjof Nansen, the plane was well below the summit of that towering block. Byrd described the scene to port as they emerged over the headreach of Liv Glacier: "A whole chain of mountains began to parade across the eastern horizon. How high they are I cannot say, but surely many of them must be in excess of 15,000 feet, to stand so boldly above the rim of the 10,000 foot plateau. Peak on peak, ridge on ridge, draped in snow garments which brilliantly reflected the sun, they extended in a solid array to the southeast." These were the mountains mapped by Amundsen as trending in a straight line to the southeast, affirmed now by aerial observation, although they appeared not to extend beyond 87° S. McKinley photographed an overlapping set of oblique shots of these mountains from which triangulations were later made.

Figure 5.10. In a view close to what Byrd would have seen looking out the right side of his plane on the return leg from the South Pole, Amundsen Glacier funnels into a steep defile as it cuts its way to the ice shelf. The northern end of Nilsen Plateau appears from the right, sending down a branching system of ridgelines and spurs. On the left edge of the glacier is Breyer Mesa, the site of one of the survey stations established by the Topo East party in 1963 (Chapter 7). At the right rear, the Watson Escarpment peeks beneath a layer of cloud on the far side of Scott Glacier. The ghostly summit standing above the left skyline is Mount Goodale (see Figs. 5.6, 6.2).

From the vantage of the plane McKinley also photographed a major outlet glacier unrecognized by Amundsen that funneled through the mountains on the near side of Mount Thorvald Nilsen (Fig. 5.10). Byrd eventually named this Amundsen Glacier. Above the spot where the glacier drops steeply into the mountains is a broad, semicircular area riddled with crevasses that feed the glacier. The irony is that the Devil's Glacier, so named by Amundsen, is in fact a part of the headreach of Amundsen Glacier itself.

Far out to the southeast Byrd noted "what appeared to be the largest glacier that we had yet seen, discharging into the new range we had first observed on the base-laying flight." This was the upper reaches of Scott Glacier, observed for the first time. Looking back at the backside of the mountains, all but topped by the massive ice sheet, Byrd shivered at the sight of "a line of low-hung peaks standing above the swelling folds of the plateau. Now, with the full panorama before us, in all its appalling ruggedness and gothic massiveness, we had a conception of the ice age in its full tide."

At 1:14 A.M. on November 29, 1929, the plane arrived at the South Pole, crisscrossed it, dropped some flags, and headed back to Little America via Axel Heiberg Glacier. McKinley continued photographing out the left side of the plane as they dropped between Mount Fridtjof Nansen and Mount Don Pedro Christophersen, and then landed at the fuel depot at the mouth of Liv Glacier. Once the fuel was pumped, the Ford trimotor flew straight back to Little America, to a hero's welcome but without the polar feed.

With the plane safely returned, the geological party was eager to tackle the mountains. The peaks looked so close that Gould boasted, "We'll make the mountains today or bust." And bust they almost did. First, the distance was much greater than had been estimated, a common problem in Antarctica, where scale is absent. Then the party was drawn into a treacherous crevasse field, whose one saving grace was that it was in blue ice, so the crevasses were obvious from their white bridges.

The crevasse field at the eastern portal to Liv Glacier was analogous to, but not nearly as chaotic as, other junctures along the Transantarctic Mountains, where outlet glaciers enter the ice shelf. At first, large swells rose and fell, perceptible only when two of the sled teams disappeared from view. A few narrow crevasses signaled the edge of the field, and for a while the going wasn't bad. The blue ice made it difficult for the men to maneuver the sleds and for the dogs to gain traction, and merely knowing where the crevasses were did not always mean that the party could avoid crossing them. The deeper into the field they plunged, the more hazardous the obstacles became. At the worst the men let the dogs pull the sledges ahead and were towed along behind by ropes. Bridges dropped repeatedly at the men's heels but never quite took them down.

A route slightly to the east would have avoided the problem, but the mountains were such a single focus that by the time the men realized their position, it seemed too great a risk to retrace their steps. The unknown that lay ahead at least offered hope that the situation would improve. When they finally began to come out the far side of the crevasse field, they stopped on the first snow surface large enough to tie the dogs and set up camp, less than a mile from the nearest outcropping of rock. The meters on the sleds recorded an incredible forty miles that day, twenty of which had been in the thicket of crevasses.

The next morning Gould awoke before the rest of the party, drawn to the rocks as though the Sirens were singing there. Quietly he dressed, put on skis, and alone started for the nearest exposure. The glacier surface here was snow covered, concealing the crevasses that lay in his path. Within a few hundred yards a bridge collapsed beneath him. He pitched forward, catching himself by the arms at the edge of the crevasse. A desperate struggle brought him over the lip to safety. Shaken by such a near miss, Gould sheepishly retraced his tracks to camp and quietly crawled back into his bag, where he waited for the others to rouse.

From the reports of the geologists who preceded him, Gould knew the geology of the Transantarctic Mountains as far as Beardmore Glacier, a basement of old igneous and metamorphic rocks, covered by the Beacon sedimentary rocks intruded by sills of dolerite. From the approach, the mountains looked as if they fit this pattern. What Gould wanted most to do was find a fossil, something that would tell him the age of the rocks, and link its story to those stories of other synchronous sequences scattered around the world. Shackleton's samples of Cambrian *Archaeocyatha* from the Beardmore Glacier area actually were from the basement in rock that was unmetamorphosed, but from the looks of it, the basement rocks in this region were too metamorphosed to contain fossils. The Beacon sequence had until then produced two sets of fossils: the fish plates discovered in the lower layers of the section dating from the Devonian, and the *Glossopteris* plants from higher in the section, representing the Permian. A great deal of geologic time was unaccounted for between these two fossil levels. Plus, beds above potentially could have fossils younger than the Permian. Would the layers on the slopes of Mount Fridtjof Nansen hold this gift of time for Gould?

Mount Fridtjof Nansen towers at the apex of a triangle whose sides are Liv and Axel Heiberg Glaciers (see Fig. 5.6). The Liv empties northward, the Axel Heiberg to the east, and the rangefront strikes northwest-southeast between the glaciers. Gould figured that Liv Glacier would give ready access to the slopes of Mount Fridtjof Nansen, so after breakfast he, Thorne, and O'Brien strapped on crampons and started up its eastern side. When the others saw the tracks that led into the open crevasse, they guessed that they were Gould's, and he took a sound roasting as they proceeded. Again distances were deceiving. A nunatak in midstream that had been their goal was found the next day, after triangulations, to be twenty-one miles away. The crevasses became increasingly hazardous, eventually forcing the party to turn back. A try later that day at crossing Liv Glacier met with another underestimate of distance and even worse crevassing. Thwarted so soundly by the Liv Glacier, Gould collected some schists from the nearby spur and decided that it would be better to try some access to the southeast, perhaps following Amundsen's route up Axel Heiberg Glacier.

On their way the party wasted a day and a half in a vain search for the depot left by the earlier flight. Despite carefully triangulated air photos that had been dropped to the party, the cache remained lost in the swells north of the mouth of Liv Glacier. After lunch on December 3, the men finally headed east in close along the rocks that descended as a series of spurs from a crestline parallel to the coast (Fig. 5.11; see Fig. 5.9). All of the rocks were of the basement, no sandstones here for Gould, but the outcrops held not only schists as seen at the Liv camp, but also small intrusions of granite, which produced spectacular contact relationships with the metamorphic rocks (Fig. 5.12).

By evening the party had reached the mouth of a small glacier that drains the face of Mount Fridtjof Nansen. As a route to the base of Mount Fridtjof Nansen, its surface looked snowy and promising, so here they made a depot for the return journey and killed all but twenty-one of the dogs. Vaughan insisted that it was his duty to dispatch them, and none of the others argued. This base was called Strom Camp, after Sverre Strom, the carpenter on the expedition who, along with Bernt Balchen, had built the sleds that were used by the geological party. Later the glacier also came to bear his name.

On December 6 the party started up Strom Glacier (see Figure 5.4). At first the ascent was up and down over broad undulations, then crevasses became a hazard, but by the end of the day they were camped three thousand feet higher and fifteen miles farther on. The next day, what appeared to be an easy climb to the rocks was protracted in a tedium of crevasses, so that when the men and dogs reached a relatively crevasse-free level, they pitched camp, only four miles past and eighteen hundred feet above their last.

Finally, on the morning of December 8, the sedimentary section on the massif seemed to be in reach. Gould, Thorne, Crockett, and Goodale fastened on their skis and roped up to begin an ascent to one of the steep spurs reaching down from the face of Mount Fridtjof Nansen. As typically happens, the closer they approached the rock, the steeper and more hard-packed the snow became. Eventually the men were sidehilling with their skis to gain the necessary altitude. When at last they reached a saddle on the ridge, the snow carried on another eighth of a mile up the crest. The closer to the rock Gould climbed, the more his spirits fell, for the sandstone he expected to find appeared increasingly to resemble volcanic lava flows, sure to contain no fossils. So it was with doubled excitement that he did indeed collect samples from an ancient, sandy river on the ridge of Fridtjof Nansen late that day.

In a radio message that evening to Byrd, Gould's enthusiasm flowed: "No symphony I have ever heard, no work of art before which I have stood in awe ever gave me quite the thrill that I had when I reached out after that strenuous climb and picked up a piece of rock to find it sandstone. It was just the rock I had come all the way to the Antarctic to find."

Figure 5.11. As Gould's party mushed east from the mouth of Liv Glacier, it passed this row of granite buttresses at the eastern end of the Duncan Mountains.

Figure 5.12. At this locality, about a mile to the west of Fig. 5.11, Gould noted this spectacular outcrop with its black rafts of metamorphic rock shot through by white veins, floated in multiple stages of gray intrusion, then frozen in time.

The men were so engrossed in the rocks they examined that they failed to notice a fog collecting until it had settled in on them. Hurriedly putting on their skis and roping up, the four started to ease their way down the snow slope in nearly complete whiteout conditions. Mike Thorne, the only good skier in the group, kept order while the others slipped and fell. They passed above an open-crevassed area and then were safe to glide on down to their tents.

The next day the party moved camp another six miles up Strom Glacier, this time stopping very close to rock on another of the spurs framing Mount Fridtjof Nansen. This spur gave Gould better access, and all afternoon he climbed up through the layers, measuring more than two thousand feet of vertical section, observing the changes at different levels, and collecting samples for return and further study. Near the bottom the rocks were yellow sandstone, marked with many troughs and dipping sets of sand, the sort laid down in the channels of active rivers. Higher still were layers of black mudstone, from flood plains and swamps beyond the rivers' banks. At the top of the climb Gould discovered some black seams so rich in organic matter that they classed as low-rank coal. Smoldering in the flame of a lighted match, a small lump filled his nostrils with the incense of a peat swamp several hundred million years out of the past. Exciting as the discovery of coal was, the rocks refused to yield any fossils, so the age of the layers could be guessed at only by comparison with rocks in southern Victoria Land.

The following day the party descended Strom Glacier, detouring to the east to climb the nunatak that they took to be Mount Betty. Locating Amundsen's site would permit the geographical observations of the two parties to be tied together, but no cairn could be found on the top of the small peak, so the party returned to Strom Camp feeling frustrated. One last attempt was made on December 12 to locate the depot of the polar flight, and this effort succeeded. The triangulations had been put on the wrong air photo. Five gallons of gasoline and two hundred pounds of man food were recovered, giving the party a safety net as they headed east into the unknown in search of Carmen Land.

Three weeks of provisions were lashed to four sledges, the remainder was depoted at the Strom Camp, and although it was Friday the 13th, the party energetically set a trail to the mouth of Axel Heiberg Glacier, passing the northern end of the range that Amundsen's party had climbed when they last passed through. A little inside the western portal of the glacier, Gould wrote, "No camp of our whole journey was in a more complete fairyland setting," as he looked at the Axel Heiberg flooded in low, golden sunlight, pouring between the snowbound massifs of Christophersen and Nansen (see Fig. 5.5). The next morning, however, when the party awoke, the camp was engulfed in a windless cloud that was dropping soft snow. This total whiteout condition lasted for three days, as temperatures at one point reached 35° F. Water poured through the tents, wetting everything. The dogs were wet and miserable.

At last, on the morning of December 17, the cloud pulled back onto the ice shelf and the soggy party headed out, with Thorne and Gould on skis breaking a trail in the deep snow for the dogs to follow. By midmorning they were halfway across Axel Heiberg Glacier, where a snow cairn was built for Thorne's triangulations of peaks around the catchment. Crevasses ahead forced the party to turn out toward the ice shelf before heading east again. The 18th was a fine, clear day with a number of stops to take photographs of

Figure 5.13. Swaths of crevasses roughen the surface of Amundsen Glacier. Snaking into view from behind Breyer Mesa, this torrent of ice sweeps under Nilsen Plateau, then caroms through staggered ridgelines to the ice shelf in the fore. There Gould's party was passing on the afternoon of December 17, 1929.

the mountains and glaciers that were opening to the south. To the east of Mount Ruth Gade was a snowy glacier cutting a straight and narrow swath through the mountains from the plateau. This was named Bowman Glacier for Isaiah Bowman, director of the American Geographical Society (see Fig. 5.3).

By midday the party was "looking up the most stupendous glacier" that they had yet seen, exceeding even the Liv in scale. Gould named this Amundsen Glacier, a ribbon of lacerated ice flowing down from the ice cap to the west of the bold, flat-topped Nilsen Plateau, through a middle and foreground (Fig. 5.13). Here the sedimentary layers on the upper slopes that gave the mountains their blocky character retreated far back from the mountain front, as the foreground was dominated by jagged, alpine peaks composed of granite and metamorphic rocks of the basement. Three peaks that rose high above all the others at the eastern boundary of Amundsen Glacier's catchment were named Mounts Crockett, Vaughan, and Goodale (see Fig. 5.6).

Later in the afternoon, a vast field of blue ice loomed ahead of the party, so the men turned abruptly south to the closest outcrop of rock, where they camped with the hope of gaining a vantage of what lay ahead. These foothills were rounded nubbins that had been flowed over in the past by ice at a higher stand of the glaciers. Their bedrock was composed of metamorphic layers interspersed with granitic dikes. A novelty of these rocks was the presence of green stains indicating a low level of copper mineralization. Gould named the summit O'Brien Peak, for the surveyor of the party. The view to the east showed no way to avoid the blue ice and to continue in that direction. Moreover, the growing doubts about the reality of Carmen Land were finally laid to rest (Fig. 5.14). Gould states, "We found that the range did not trend southeastward from the vicinity of Axel Heiberg Glacier as Amundsen thought but rather it is continued in an almost due easterly direction from it. Furthermore we looked away toward the east and north for

Figure 5.14. In this view to the northeast from the crest of O'Brien Peak (see Figs. 5.6, 6.2), Gould was finally able to verify that Carmen Land was a figment without form. The perfectly flat surface of the ice shelf marked the horizon. In this image the horizon is darkened because of a bank of cloud shadowing it in the distance.

signs of Carmen Land. There were none. We were now sure beyond any shadow of doubt that it did not exist."

The next day the geological party headed east into the icefield. The surface was ablated into sharp cups that made skis useless, so the men rode on the sledges. Crevasses were numerous, but they were small and fairly obvious in the blue ice. The dogs slipped a lot, and the sleds took a beating. The men were so absorbed in their movement that they failed to watch what the mountains were doing until they stopped to make camp for the night. The rangefront had receded into a great cleft that drained ice from the polar plateau. The path was due south and nearly straight, and it appeared (correctly so) that nothing lay between the beholders and the South Pole (see Fig. 6.1).

The field party was camped right in the middle of the mouth of a massive outlet glacier. Gould wrote of the place, "My feelings were a mixture of curiosity and very real awe, as I looked up this new glacier and across the mountains to the east, knowing that ours were the first human eyes ever to look upon them." Mounts Goodale, Vaughan, and Crockett were arrayed along the western margin of its catchment. As with Amundsen Glacier, jagged peaks lined the middle reaches, and far in the distance flat-topped massifs bounded the plateau. But here the scale was nearly twice that of Amundsen Glacier, both in length and width. Thorne Glacier is the name that appears on the maps from this expedition, but some years later the name was changed to Scott Glacier, the rationale being that major outlet glaciers should bear the names of major expedition leaders.

One day more and the party stopped at their farthest point, an arching ridge several miles long composed of schist (Fig. 5.15). This they named Supporting Party Mountain to honor the men and dogs who had laid the depots that had enabled them to reach so far.

Figure 5.15. Supporting Party Mountain as seen from the south (see Figs. 5.6, 6.2). Both Gould's and Blackburn's parties camped in the valley between the two dark ridges at the right edge of the image. The west (right) side of the mountain has two ridge-lines overlapping in this perspective. The ascent route up the ridge to the rear is marked, but the upper reaches of the route are out of sight behind the summit spotted by the dot. The prominent massif on the horizon to the left of Supporting Party Mountain is Mount Griffith, forty-five miles distant across Scott Glacier. To its left with face in partial sun is Mount Vaughan.

Freddy Crockett drew the short straw, so while he tended the dogs, Gould and the others climbed a side spur to where it joined the high point on the ridge crest.

To the southeast the mountains were compressed to a single cliff face dropping three thousand feet from the ice cap to a terrace thirty miles across, studded with ridges and peaks in low relief. The horizontal top of the escarpment, bare of any sedimentary layers, appeared to descend toward the horizon in the east. The ancient erosion surface embedded in the faces of Mount Fridtjof Nansen and many peaks beyond was here exhumed, a plain once more (Fig. 5.16). In its former existence it had been a lowlands cut and drained by swift rivers. Reincarnated now, the plain had an edge of cliff backed by a fringe of ice thickening toward the south.

To the south and branching to the west was a stately range culminating in a pair of faceted peaks (Fig. 5.17). Byrd later named the twin summits Mount Gould. In front of the escarpment running west-northwesterly was a broad glacier that Gould named the Leverett, in honor of Frank Leverett, a Quaternary geologist with the U.S. Geological Survey who had inspired Gould's interest in glacial geology when teaching the undergraduate in a course at the University of Michigan. Although a view of its headreaches was blocked by a ridge in the middle ground, it appeared that the glacier might extend many miles to the east, causing Gould to speculate that Leverett Glacier could be longer than either the Amundsen or Scott, and could perhaps contribute a volume of ice to the ice shelf second only to the Beardmore.

At the top of Supporting Party Mountain the men built a cairn and left the following note:

December 21st
Camp Francis Dana Coman
82° 5′ 7″ South
147° 55′ West
Marie Byrd Land, Antarctica

This note indicates the furthest east point reached by the Geological Party of the Byrd Antarctic Expedition. We are beyond or east of the 150th meridian and therefore in the name of Commander Richard Evelyn Byrd claim this as a part of the

Figure 5.16. From the summit of Supporting Party Mountain, Gould's party looked to the south and east, where the mountains narrowed to a single cliff, disappearing from sight far to the east. A broad glacier drained the front of this escarpment, appearing to head at its distant termination. In fact, Leverett Glacier drops through a narrow cleft in the face of the Watson Escarpment, not visible from Supporting Party Mountain (see Fig. 5.6). In this image from Mount Webster, ten miles to the southeast of Supporting Party Mountain, Evans Butte, the flat-topped outlier of the exhumed erosion surface on Watson Escarpment, deflects cold, gravity-driven, katabatic winds into lenticular clouds. Watson Escarpment, mantled in its own ground-hugging katabatic cloud, appears in the left distance and peeks through the two saddles to the right of Evans Butte. Leverett Glacier flows from left to right in the foreground of the image.

FINDING CAIRNS

Anyone who travels to deep-field Antarctica experiences setting foot in places where no human previously has traveled. Even if several field parties have worked in an area, there are still many ridgelines, moraines, tributary glaciers, and drifts that are virgin surfaces with no footprints. I have always felt a certain thrill, along with privilege, good fortune, and satisfaction, when I see a new perspective and frame a camera shot of a new location. I have climbed many mountains in places other than Antarctica and have often wondered if I was the first to reach some lonely spot. But I could never be confident that some old prospector searching for gold or a herder chasing goats into high country had not previously climbed a ridge on which I was standing. In most of the Transantarctic Mountains, however, the travelers have been precious few, and those who have gone before have typically left maps and scientific reports of exactly where they went.

Nevertheless, I must admit that after thirteen expeditions into deep field Antarctica, the excitement has faded somewhat. Now, what gives me an even greater thrill is the knowledge that I am standing exactly where members of a previous field party have been and I am gazing over the same vista. I imagine their approach and wonder how it felt to them to be the first. I feel a connection, especially when I discover a cairn left by a field party as a marker of its achievement, and sometimes even a written record of who the men were and what they did.

For instance, in the early seasons after the International Geophysical Year in the late 1950s and early 1960s, when topographers were scouring the Transantarctic Mountains to prepare their maps, they built cairns at many of their survey stations. Perhaps a half-dozen times over the years, I have come upon these robust, chest-high constructions built from whatever stone the bedrock offered. Invariably they are placed at some high point overlooking a spacious panorama. Some might say that they are blemishes on pristine wilderness, but to me any cairn is an apt monument to the human history of this frozen land.

The most memorable cairn I found was left by the Gould party on Supporting Party Mountain in 1929 (Fig. S.10). During the 1977–1978 field season, my party was working in the area to the north of Leverett Glacier. From Gould's writings, we knew that his party had built a cairn at the summit of the mountain there at the easternmost reach of the exploration. In it the men had left a note claiming the territory to the east of longitude 150 in the name of the United States of America. From Blackburn's writings, we knew that his party had visited the cairn at the beginning of its traverse up Scott Glacier, had left a note of its own, copied Gould's note, and carried the original back to him. I was determined not to miss this remnant of the heroic era, and planned a day of mapping that would

Figure S.10. A modest cairn built of schist marks the summit of Supporting Party Mountain. Gould's party constructed the landmark and buried in it an unmarked can, an empty Quaker Oats tin, a broken thermometer, a bamboo splint, and a note claiming Marie Byrd Land for the United States of America. When Blackburn reached the spot in 1934, he wrote a note of his own, copied the previous note, and returned the original to Gould.

include a climb to the summit of Supporting Party Mountain and to its cairn. We approached up the gentle north ridge, whereas Gould's and Blackburn's parties had climbed up the steep western spur. Because of the convexity of the ridgeline we did not see the cairn until we were almost on it. Then there it was—an alien sign of humanity in a lifeless landscape! We peered into the chinks between the rocks and spotted the treasure deep within.

The proper construction of a cairn in which a record is left includes a stone that can be withdrawn so that the contents can be easily accessed, and indeed that was how this cairn was made. I carefully removed the doorway stone, reached in, and took out a colorful tin can that had once held dried oats for Gould's party. Inside the tin were the penciled notes, written with precision and flare in a surprisingly steady hand, given that Blackburn must have written them either with a gloved hand or a bare hand stiffened by the cold. In addition, there were a bamboo splint and a broken thermometer, left as relics of their heroic traverse.

My group lingered by the cairn, looked out at the scene that Gould and Blackburn had beheld, discussed the route that they had taken to this spot, took photos as souvenirs, then replaced the notes, added our own, and descended the mountain.

United States of America. We are not only the first Americans but the first individuals of any sort to set foot on American soil in the Antarctic. This extended sledging journey from Little America has been made possible by the supportive work of the Supporting Party composed of Arthur Walden, leader, Chr. Braathen, Jack Bursey and Joe deGanahl. Our Geological Party is composed of

L. M. Gould, leader and geologist
N. D. Vaughan, dog driver
G. A. Thorne, topographer
E. E. Goodale, dog driver
F. E. Crockett, dog driver and radio operator
J. S. O'Brien, civil engineer

From Supporting Party Mountain the geological party retraced its tracks to the Strom Camp depot. On Christmas Day, returning to the small peak that they had previously thought to be Mount Betty, the men searched in vain for Amundsen's cairn. An insignificant shoulder of rock projecting northward about one thousand feet below the summit was the only other possibility (see Fig. 5.4). When viewed through binoculars, the outcrop did seem to betray a speck of something. And to be sure, when Gould and Goodale skied over to check, they found the cairn. Although the can was a little rusted, the paraffin that Amundsen had left was still good, as were the matches. In addition, there were two notes, one giving the names and addresses of Wisting and Johanssen, who had helped build the cairn, and another torn from Amundsen's notebook stating his attainment of the South Pole. Gould took the matches and Amundsen's note and left a note of his own. Ultimately Gould gave Amundsen's note to the Norwegian Geographical Society.

Figure 5.17. No explorer has yet set foot on Mount Gould, the twin-peaked citadel rising to the south of Leverett Glacier. This image is taken from Mount Webster.

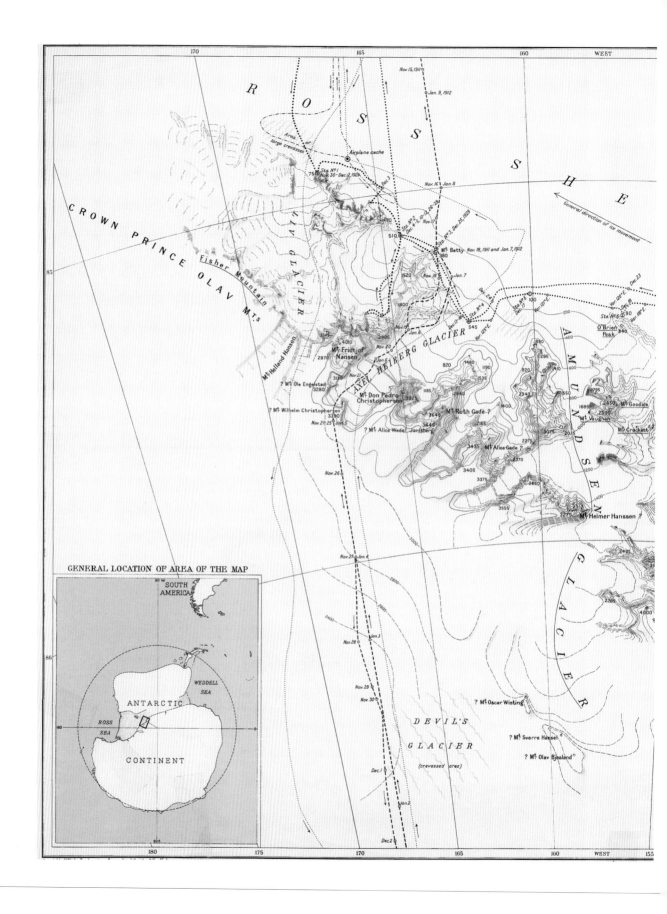

R O S S

S H

S E

Nov. 15, 1911

Jan. 9, 1912

Area of large crevasses

Airplane cache

Sta. Nº 1
75⊙ Nov. 30–Dec. 2, 1929

Nov. 16 Jan. 8

Dec. 3

General direction of ice movement

CROWN PRINCE OLAV MTS

Sta. Nº 5 Dec. 1, 26–29.

470

Sta. Nº 2
Dec. 4, 5, 10–12, 26–29.
Nov. 17

Sta. Nº 3, Dec. 25, 1929

85

Fisher Mountain

510⊡

Mt Betty Nov. 18, 1911 and Jan. 7, 1912
380

1520 Nov. 18 Jan. 7

Dec. 21

Dec. 24

Sta. Nº 4

1800

545 Nºr 125° E.

Sta. Nº 6
Dec. 24, 1929 Nºr 127° E.

Sta. Nº 7 100 Nºr 127° E.

Nºr 120° Dec. 23

Nºr 120° Dec. 18

Sta. Nº 8 90 Nºr 118° E.

200

O'Brien
Peak 840

400

Mt Helland Hansen

3245

4010

2905
Mt Fridtjof
Nansen
2970

Jan. 6.
Nov. 19

Nov. 20

AXEL HEIBERG GLACIER

820

1480 1190

1525

1295

600

800

1800

Mt Goodale

2450

3180

? Mt Ole Engelstad
3280

Mt Don Pedro
Christophersen
3925

1185

2640

2340

1650

1785

2500
Mt Vaughan

3075

Mt Crockett

? Mt Wilhelm Christophersen
3390

Nov. 21–25 Jan. 5

3645

3445

Mt Ruth Gade?

2165

A M U N D S E N

? Mt Alice Wedel Jarlsberg

2275

3435

Mt Alice Gade ?

2370

3405

3375

2650

3555

3275 Mt Helmer Hanssen ?

Nov. 26

2435

G L A C I E R

2765

4000

Nov. 27 Jan. 4

1000

2800

86

2400

Nov. 28

Jan. 3

2600

Nov. 29

Nov. 30

? Mt Oscar Wisting

D E V I L ' S

G L A C I E R

? Mt Sverre Hassel

(crevassed area)

? Mt Olav Bjaaland

Dec. 1

Jan. 2

Dec. 2

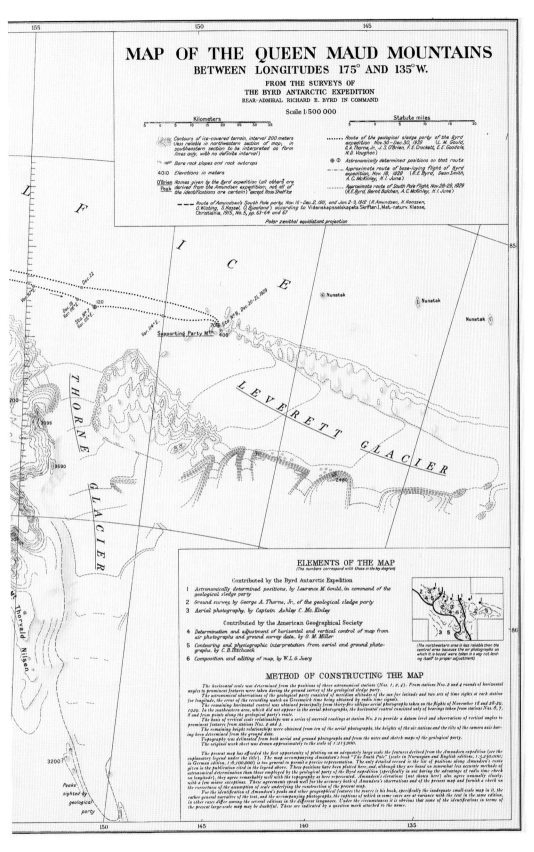

Figure 5.18. Combining Amundsen's vague mapping, aerial photography from Byrd's polar flight, and his party's survey along the mountain front, Gould published this topographic map of the Queen Maud Mountains in 1931. Compare with Fig. 5.6. Reproduced by permission of the American Geographical Society.

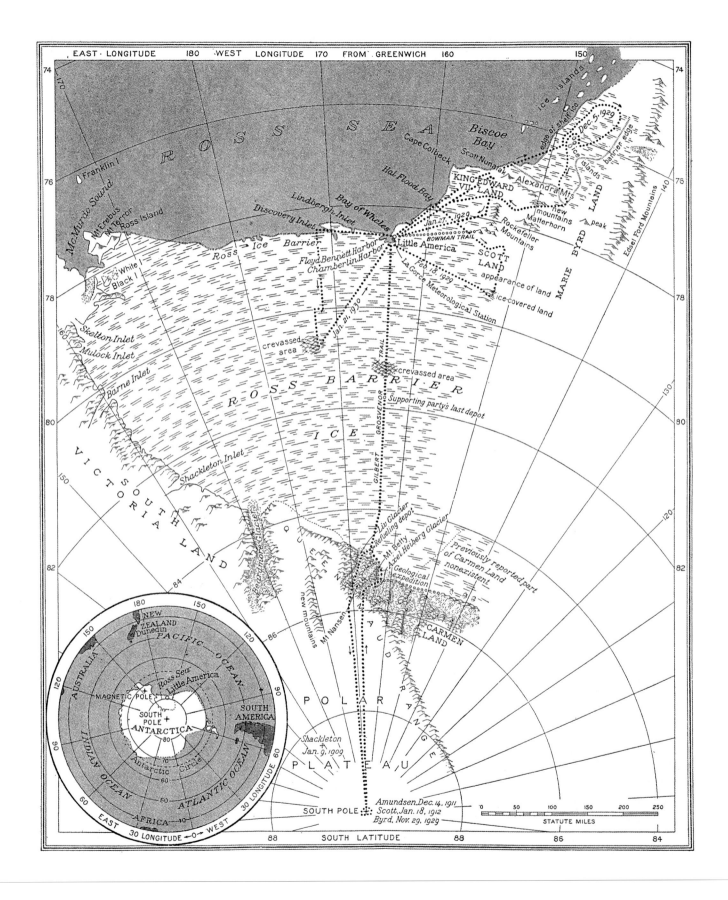

Figure 5.20. Gould's view to the east of Supporting Party Mountain gave the appearance that the Watson Escarpment gradually dwindled in height for more than 50 miles into the headreaches of Leverett Glacier. In fact, Leverett Glacier turns abruptly south about 30 miles to the east and cuts a cleft into the face of the escarpment (see Fig. 5.6). Twenty miles beyond this, what Gould saw as a dwindling escarpment is in fact a major ridge system that drops to the east-northeast from the plateau. Beyond this ridge the Transantarctic Mountains extend for another 150 miles. This image, which reaches nearly to the mountains' end, is taken from the 10,400-foot summit of Mount Analogue, at the crest of the ridge system to the east-northeast of the escarpment. The 6,000-foot southern face of Mount Doumani dominates the foreground. The dark massif in the middle distance is Mount Teller. Beyond is Reedy Glacier, the last of the outlet glaciers, which slips from right to left in front of the dark mountains at the center skyline. Off at the deep end of the horizon, the thin edge of the Horlick Mountains disappears from sight beneath the ice.

Figure 5.19. (opposite) Byrd's map of discoveries, published in 1930. Note the repositioning of Carmen Land, which was dropped from Gould's map a year later (compare Fig. 5.8). Reproduced by permission of the Richard E. Byrd family.

On December 31, after repacking and giving the dogs a rest, the party set off across the Ross Ice Shelf for Little America. Twenty days later the men arrived at their destination. In the course of their journey, 1,525 miles had passed beneath their runners.

Perhaps more important than geographical discoveries, BAE I demonstrated the possibilities of airborne exploration into the heart of Antarctica, traversing territory in a few hours that would occupy a ground party for an entire season. Aerial photography also proved its worth for cartography, but ground truth for accurately recording elevations remained the domain of the surveyor. Gould's party provided that control, and in the course of its journey along the mountain front extended the known limit of the Transantarctic Mountains another sixty miles beyond Amundsen's sighting (Figs. 5.18, 5.19).

Although Gould speculated that the mountains terminated at about the limit of his view, the Horlick Mountains (Fig. 5.20), being the true termination of the Transantarctic Mountains 150 miles east of Supporting Party Mountain, were vaguely spotted during Byrd's Second Antarctic Expedition (BAE II) on November 22, 1934, from the endpoint of a flight from Little America to the southeast across the Ross Ice Shelf. What no one yet realized was that at about the mouth of Scott Glacier the ice shelf is grounded, and from there the West Antarctic Ice Sheet steadily rises in elevation, finally merging with its East Antarctic counterpart and overriding the mountains completely.

Finally, Gould's party demonstrated the nonexistence of Carmen Land. As the Parry Mountains had been to Ross, so was Carmen Land to Amundsen. But who was Carmen, the namesake of the phantasm? Amundsen never said, and historians have yet to determine. Was she the fiery gypsy of Bizet's opera, conjured to counter the gripping cold of the polar march, or perhaps a mysterious woman of Madeira who had shared her warmth with Amundsen on that last night in port? All that may be said is that whoever she was, she remains as ephemeral as the land that was given her name.

6 Earth's Land's End

The Exploration of Scott Glacier

The First Byrd Antarctic Expedition had been extremely successful at capturing the imagination of the American public. So in spite of a homecoming to the Great Depression, Byrd sought to return to the Ice and extend the discoveries of his recent campaign. The creation of the Second Byrd Antarctic Expedition (BAE II) at a time of such national austerity is largely to the credit of its leader's exceptional promotional skills. With the donation of materials and supplies from many manufacturers and the public donation of funds, the expedition sailed south in autumn 1933, one year delayed from the targeted departure.

Byrd had brought the age of airborne exploration to Antarctica, and on his second venture he intended to push the horizon even farther from aloft. The pride of the expedition was a new two-motor Curtiss-Wright Condor biplane equipped with both skis and floats, which had a range of thirteen hundred miles fully loaded. As the ship approached the pack ice, the plane was lowered to the water for takeoff and used to scout leads through the pack ahead of the vessel. On January 17, 1934, a landing party reoccupied Little America at the Bay of Whales.

Even more than its predecessor, the Second Byrd Antarctic Expedition went forth in the name of science. Among the fifty-six men who wintered over were two geologists, a geophysicist, three physicists, two meteorologists, three biologists, a physician, a surveyor, and an aerial photographer. Numerous technological advances were brought to the service of the expedition. Whereas the public had been informed of the progress of Byrd's first expedition through messages radioed via teletype, voice broadcasts through the Columbia Broadcasting System were a weekly feature of the second, making "our men in Antarctica" part of the daily conversation.

A major goal of the expedition was to determine the thickness of the ice sheet and to ascertain whether or not a water passage linked the Ross and Weddell Seas. Techniques developed for the exploration of oil, employing reflected seismic waves produced by explosions at the surface, were used to map the ice-rock interface at the base of the ice sheet.

Because of the ambitious air agenda, weather forecasting was an integral part of the program. The expedition continuously recorded systematic weather observations for 365 days at Little America, and all remote parties kept weather logs. A plan to winter over three men for meteorological observations at an advanced base camp on the polar plateau had to be curtailed when depot laying was less successful than had been hoped. Finally, Byrd occupied a small station by himself one hundred miles south of Little America. Inadequate ventilation in his hut, a room nine by thirteen feet sunk into the snow of the ice shelf, created a lethal struggle for the man, balancing the carbon monoxide produced when the stove was fired against the penetrating cold when it was not. With long lapses of consciousness, Byrd's condition finally became apparent to those on the radio back at Little America, prompting a risky rescue in the polar night by tractor and airplane, returning the admiral to safety.

Also, for the first time in Antarctica, the expedition used mechanized ground travel productively. The vehicles included a Cletrac tractor, two Ford snowmobiles, and three Citroen trucks fitted with tracked treads and front skis. Successful runs south on the ice shelf during the austral autumn of 1934 established a series of depots 50, 100, 125, and 155 miles from Little America. Dog teams also served the expedition, but the machines proved themselves able to transport heavier loads over greater distances for longer periods of time.

Plans for the following summer were to send a geological party eastward to explore Marie Byrd Land, to send a second geological party south to Supporting Party Mountain, there to proceed to the east beyond Gould's farthest point, and to send a geophysical party south to ascend Scott Glacier and attempt to measure the thickness of the ice of the polar plateau. The southern parties were to use dogs for their travel, but the Citroen tracked vehicles supplied depots out 300 miles. The greatest uncertainty was whether the tractors could make it through the heavily crevassed area encountered by Amundsen's and Gould's parties at 81° S.

On November 1 the geological and geophysical parties waited at the 209-mile point while the Citroens closed on their rear. When the tracked vehicles failed to appear, the geological party returned to mile 193, where they found tractor no. 3 wedged in a gaping crevasse. With great difficulty the machine was extracted from the slot. After consulting by radio with Byrd, the geophysical party was recalled, and the combined parties returned to mile 159, where they tried another southerly route, 10 miles to the east. When this route proved equally impassable, they decided to send the geophysical party eastward with the tractors onto the West Antarctic Ice Sheet, and the geological party south with dogs, supported by two members of the geophysical party. On November 14 at the 300-mile point, they laid a depot, and the supporting party returned to Little America.

The geological party numbered three. The leader, Quin A. Blackburn, was geologist and surveyor; he was the man who had survived a night out in a storm dug into a hole in the snow with his huskies during the first Byrd expedition. The others, newcomers to the Antarctic, had enlisted as dog drivers: Stuart D. Paine, navigator and radio operator, and Richard S. Russell, Jr., who was in charge of transport and supplies. From the 300-mile depot they drove three thirteen-dog teams with 1,400 pounds of dog food and 450 pounds of human food. During the next afternoon, well on the trail, the men caught their first glimpse of the Queen Maud Mountains far ahead, a gray horizontal that was both their goal and their starting point. That night they also shot their first dog, one of Russell's team, and cached him for the return.

For the following two days a blizzard with sixty-knot gusts pinned the men in their tent. Then on November 18 the party was nearly eaten alive by a crevasse field. First Russell's sled wedged crosswise in a crevasse hanging by one runner, then seven of Paine's huskies went headfirst into a crevasse hanging by their harnesses in space. In the worst incident, both of Blackburn's sledges went into a crevasse. The three teams were tied in train with rope, with Blackburn's at the rear. When his sleds went down, the dogs went to their bellies, flat out clawing, and Paine jammed a tethering pole into the snow to stop the slide. Blackburn had caught himself and managed to climb out of the crevasse. All of the navigational and geological gear plus a portion of rations and clothing swung in the yawning gap. A seven-hour operation ensued, during which Blackburn and Russell alternated at descending into the crevasse on the end of a rope to send the dismantled load up a line. Making matters worse, this all occurred under a stiff wind streaming past from the southeast.

Meanwhile, the Condor had started taking exploratory flights from Little America whenever clear weather permitted. Several of these sorties were into unknown territory along coastal Marie Byrd Land, but at 12:15 A.M. on November 22 the fully loaded biplane took off in a southeasterly direction, to explore the continuation of the Transantarctic Mountains out beyond the farthest sighting of Gould's party in 1930. The geological party, about fifty miles out from the mountain front, made hourly weather reports throughout the early hours of the day.

After crossing a succession of crevasse fields, the Condor continued out onto the featureless surface of the West Antarctic Ice Sheet. A little before 6:00 A.M. the plane was reaching the limits of its fuel, but because the weather conditions were ideal, the pilot decided to push forward for a few more minutes. Then "a cluster of snow-clad peaks" materialized on the horizon, perhaps a hundred miles off. They were too distant for the photographic camera to be of use, but binoculars brought them in with certainty as the plane held its course to fix the bearing of these distant mountains. The estimate was that the peaks were located at approximately 85° 30′ S and between 110° and 115° W, about 170 miles beyond Gould's sighting. This distant protrusion was called the Horlick Mountains, after William Horlick, one of the major contributors to the expedition. Were these mountains a continuation of the Queen Maud Mountains, or were they some lonesome range set in extreme isolation? The answer would not be known until the end of the decade, when Byrd again would lead an expedition to the icy continent.

After issuing weather reports, Blackburn's party slept for the day and then packed camp and pushed forward until 11:00 A.M. the following morning, running an exceptional thirty-three miles, stopping twelve miles out from Supporting Party Mountain. At that camp they cached supplies in a depot for the return and marked out a one-mile-square area as an emergency airfield. The party had been regularly killing dogs and either caching them or eating them. When they arrived at "Mountain Base Depot" they had twenty-seven, and there shot another nine for the return.

At last the mountains were at hand. Looking off to the east, the south, and the west, Paine recorded, the "grand panorama of black faces, jagged peaks, bold contours of a continent submerged by the rigors of a glacial climate, and all subdued by a mellow purple haze. Sometimes it is more Maxfield Parrish than the artist himself. But we must get to them."

Another blizzard pinned them down until November 26, when they were able to move in to the base of Supporting Party Mountain, the farthest reach of Gould's party five years before. From here on the party had two teams of nine dogs, with Blackburn skiing out in front looking for crevasses, followed by the dog drivers—first Paine who navigated the traverse, followed by Russell. The next morning Blackburn was up early collecting specimens of rock on the spur behind camp, and taking pictures with the Zeiss and Leica. The rocks were a highly contorted black schist, a metamorphic rock with an abundance of the black mica, biotite. Red crystals of garnet and veins of white quartz were also laced through the rock. Like Gould before him, Blackburn noted the occurrence of these metamorphic rocks and collected them, but their complexities were beyond the scope of such a fast-moving, exploratory expedition, and they lacked any trace of fossils that might have been preserved in the sediments before their metamorphism, so inherently they could add little to the geological story.

Later that day all three men climbed to the summit of Supporting Party Mountain, where they found the cairn erected by Gould's party. Blackburn copied the note that Gould had written and took the original back with him, later presenting it to Gould. He also left a note of his own:

Base of Queen Maud Range
Marie Byrd Land
Antarctica
Nov. 27/34

This note is left here by the Queen Maud Geological Party of the Byrd Antarctic Expedition II. We are here to conduct a geological reconnaissance if the short time at our disposal we find a feasible way up for our dog teams. Otherwise we may proceed up Thorne Glacier or work to the east. In any case our plans call for arriving back at our mountain base 14 miles out on the Barrier on December 23 and returning then to Little America.

Our camp is situated at the N.W. base of this peak in honor of which we are calling it Supporting Party Mt. Camp. Clouds obscuring most of the view but there are indications of clearing to the N.E.

Stuart D. Paine
Quin A. Blackburn
Richard Russell

At the summit Blackburn also found faceted pebbles and striations on the bedrock indicating that all of Supporting Party Mountain had been overridden at some time in the past by glacier ice standing at least one thousand feet higher than the levels around the base of the peak today.

Stratus clouds closed in while the men were on the summit, creating beautiful shadow patterns on the snowfields below, but permitting only fleeting views of the mountains. For a time the clouds rolled back revealing Leverett Glacier banked along the Watson Escarpment, its headreaches seeming to arise somewhere far to the east beneath the cliff face, but the distant parts were obscured by a shoulder of land. With no apparent access to the plateau in that direction, and with most of the face of Watson Escarpment mantled with icefalls, Blackburn chose instead to venture up Scott Glacier, where the image of bare rock faces and tier upon tier of snow-clad peaks had slowly taken shape as the party approached the range. First the mountains had been a thin band of blue-gray at the horizon, then the main massifs took form between the outlet glaciers. Eventually the ridges had begun to reach out and down from the blocky summits. And with the sun directly out of the north, demarcations of light and dark had hinted of exposures of bedrock through the ice. Close in Scott Glacier could be seen to issue from a deep corridor lined by mountains, gabled and domed, utterly unknown, a field geologist's dream (Fig. 6.1).

The next day, November 28, the party headed west-southwest toward Durham Point at the gateway to Scott Glacier (Fig. 6.2). Where Leverett Glacier enters the Ross Ice Shelf, it is thrown into broad rolls of rippled blue ice. After crossing its mouth, the party collected some more schist at a small nunatak along the route, and then pitched camp about a half mile east of Durham Point (Fig. 6.3). After dinner Blackburn went out to the cliff face just south of camp. There the rock was a gray granite cut through with many coarsely crystalline mineral veins, called pegmatites. The potassium-bearing feldspar, microcline, gave the pegmatites a pinkish blush.

The following day temperatures were a mild 16° F, but a twenty-knot wind from the east afflicted the party. A swift run down an ice slope brought the men to the rock, where they built a cairn, marking their passage and naming the feature Durham Point, after Paine's hometown of Durham, New Hampshire. Steering wide to avoid the steep, pressured ice at the cape, the party rounded the point and began the ascent of Scott Glacier. Strato-cumulus clouds gathered and shifted over the nearground, and cumulus capped the distant peaks, with a heavy ground drift swishing past the party's knees. At clefts in the rocks to the east the wind poured through with amplified vigor.

That evening the party camped in the shadow of Mount Hamilton (see Fig. 6.3), close enough to the face to receive some protection from the continuing easterly wind. Paine recorded, "To-night had hamburger hash, milk, sherry + ice cream. A strange meal in a strange place and Thanksgiving too." The rocks at the base of the mountain again were gray granite cut by pink pegmatitic dikes, but here also was some of the black schist like that at Supporting Party Mountain. The granite clearly crosscut it, demonstrating that it was the younger of the two rocks, though the absolute age of either of the rocks was still beyond the capabilities of the technology of the day. From there on to the south Blackburn would take no samples, resuming only when the party was homeward bound

Figure 6.1. (overleaf) Tier after tier of ragged peaks line the margins of Scott Glacier, furrowing through the Queen Maud Mountains and skirting the left (east) side of Taylor Ridge, from which this image is taken. In 1929 Gould's party was the first to gaze up this magnificent defile. Blackburn's party traversed along the east side of the glacier, camping at the northern end of the Zanuck massif, the prominent feature with three summits. With cirques and small hanging glaciers halfway up its face, Mount Zanuck is the twin-peaked central summit. To the right, rising out of Scott Glacier, is Grizzly Peak, a beacon that can be seen from both ends of Scott Glacier. The summit to the left of Mount Zanuck is unnamed. At the head of the glacier the La Gorce Mountains are lost in cloud.

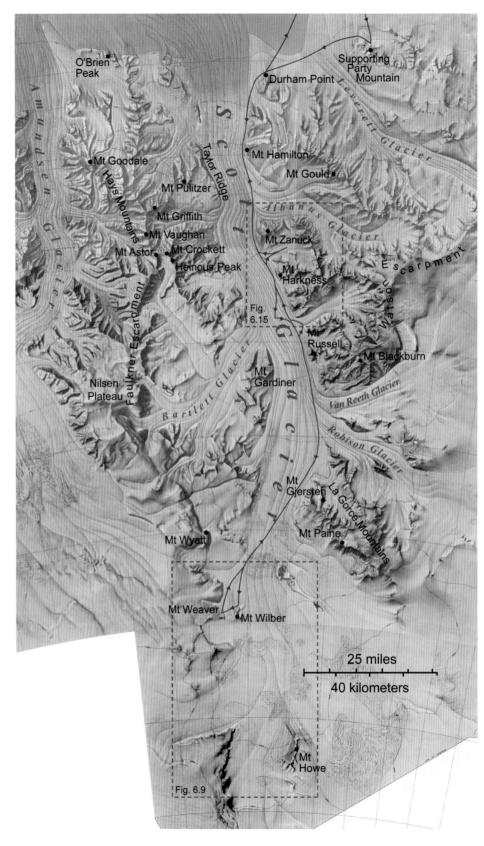

Figure 6.2. Shaded-relief map showing the route taken by Blackburn's party to the head of Scott Glacier and back. Orange dots mark the survey stations of the Topo East party during Operation Deep Freeze 63.

from its southernmost penetration. The men would travel light and, they hoped, far, but whatever route was chosen they would be backtracking over it on the return.

The next morning as Blackburn recorded weather observations, he wrote in his log, "All visible mountains stand out this A.M. in grandeur." Across Scott Glacier the two thousand–foot sheer wall of Taylor Ridge (see Figs. 6.2, 6.3), breached only by two small icefalls, directed the glacier's flow for more than ten miles. Behind this arose a complex of ridgelines and spurs that culminated at Mount Griffith, a ten thousand–foot peak and the northernmost in a line of summits that crested the Hays Mountains (Fig. 6.4). To the south they seemed to rise to even higher elevations, but the distances were too great to be sure.

Figure 6.3. Durham Point splits the drainages of Scott and Leverett Glaciers in the lower right of the image. At the start of their ascent of Scott Glacier, Blackburn's party camped about a half-mile from this point, just outside the frame to the right. In the morning the party ran down the ice incline to the point of rock and there built a prominent cairn "about 100 feet above level of barrier." From there the party sledged south along the eastern margin of the glacier, stopping to camp at the foot of Mount Hamilton, the third massif to the rear with the faceted peak at its western end. Taylor Ridge, the dark wall in the background, dikes the western bank of lower Scott Glacier. In the rear the highest summits of the Hays Mountains are lost to stratus clouds suspended in the scene.

Figure 6.4. Bare rock and ice adorn the crown of the Hays Mountains. All of the rock is granite, except for the tip of the summit of Mount Astor, the highest peak at the center, where a tiny remnant of Beacon sandstone caps an old erosion surface on the granite. The first peak to the left of Mount Astor is Mount Crockett, followed by the faceted buttress of Heinous Peak. The summit at the right horizon is Mount Vaughan. The summits between Mount Vaughan and Mount Astor are still unnamed.

Directly south of Mount Hamilton a striking granite massif with three buttressed summits stood out boldly. The mountain, rising nearly six thousand feet above the glacier at its base, contained a terrace about midway up where several hanging glaciers of blue ice issued from small cirques (see Fig. 6.1). Talus (a cone of rocks fallen from higher slopes) mantled most of the lower slopes, but the upper parts were bare and jointed crags. In his field log Blackburn simply called it Granite Mountain. The central peak was later named Mount Zanuck, after Darryl F. Zanuck, tycoon of Twentieth Century–Fox, who helped Byrd with the motion picture record of the trip.

Traversing the blue ice between Mounts Hamilton and Zanuck, the party crossed the mouth of Albanus Glacier. Because of the perspective, it appeared to the party that this glacier headed in a circular amphitheater connecting Mount Gould and Mount Zanuck, when in fact the glacier originates another fifteen miles to the east at the foot of the Watson Escarpment (see Fig. 6.2). Carried downstream in a broad arc between large pressure ridges, a wide medial moraine of granite boulders marked the confluence of Albanus and Scott Glaciers. Blackburn speculated that the Albanus was moving faster than the Scott, for the latter contained a broad depression immediately upstream of the confluence. The next night's camp was near this moraine at the foot of Mount Zanuck. After dinner the party laid out a grid of flags for surveying on their return to see whether any glacial movement was detectable.

The following day as the party broke camp, the temperature was a warm 21° F, but the wind blew at an annoying twelve to fifteen knots. The first several miles were over blue ice with hardly a crevasse, but soon the party reached snow cover, which prevailed for the rest of the day. Past the end of the Mount Zanuck massif the men gazed up into a true amphitheater, a field of pure white barely textured with sastrugi that rose steadily to a circle of jagged turrets, bathed in rich tones of brown and ochre (Fig. 6.5). What a

Figure 6.5. The Gothic Mountains encircle Sanctuary Glacier (compare with Fig. 6.15). The western face of Grizzly Peak dominates the foreground. To the immediate left, the summit of Mount Zanuck is blanketed in cloud. The two bright peaks at the back are Mount Andrews and Mount Gerdel. Pictured end on, the Organ Pipe Peaks complete the right side of the amphitheater. The men of Blackburn's party looked longingly into this basin on their outward journey, and then camped right in the middle of it on their return. Watson Escarpment peeks from behind clouds at the rear of the image.

luxury it would have been to digress into this sanctuary of granite, to spend a day and watch the sun go round, illuminating each face in turn. Perhaps on the return, if all went well. But now Scott Glacier needed climbing.

With the vantage gained by another fourteen miles of travel, Blackburn could see from the camp at the foot of Mount Harkness that south of the Hays Mountains the uplands continued as an escarpment with a rolling, level top. But what riveted his attention most through the binoculars were the horizontal layers that held up the upper portion of the wall. There were the sedimentary rocks that Gould had found to the northwest, much closer to the ice shelf there. It looked as though these beds maintained their elevation farther south, while Scott Glacier climbed ever higher toward the ice plateau. Somewhere beyond his sight these two broad planes must intersect, and that would be the place Blackburn would stop.

The temperature on the morning of December 2 had dipped to 2° F when the men awoke to find their tent and sledges drifted in. Beside Mount Harkness they left their final depot, man food, dog food, and gasoline, and with lighter loads they pushed south into the thinning air. At the next granite spur that reached down to Scott Glacier, the pressure ridge was so large that it was insurmountable, so they headed southwest out onto the glacier toward a prominent peak, Mount Gardiner, on the west side of the glacier.

This day was the most taxing of the Scott Glacier traverse. After leaving the first pressure ridge, the party followed a long slope that led to another pressure ridge in midstream. Blackburn records that beyond this the party

> picked a way through the maze around big gashes, seracs, rolls, ice holes and ridges with Stew and Dick skillfully guiding dogs and sledges until the way ahead was barred by huge pits, holes, ridges and rolls. Thence we backtracked about 4 miles, worked out to the eastward and came out onto a long undulating slope of hard, low sastrugied snow. Thence we worked up glacier toward the mountains on the east until shortly past six P.M. and camped [Fig. 6.6]. Faced a stiff down glacier wind from about 11:30 A.M. till about 4:30. Then followed a calm when the weather seemed very warm so that the icicles disappeared from our faces and we perspired.

Speaking of the crevasse fields encountered that day, Paine wrote in his diary, "A grander evidence of the overwhelming natural forces at work in this region is nowhere to be seen. It makes our petty cares + ceremonies + don'ts seem pretty trivial to the powers in motion here."

After a calm start the next day the breeze from the south sweetened again, gaining strength throughout the day. The katabatic winds that pour through Van Reeth and Robison Glaciers were veering onto the party and heading down Scott Glacier, so that as

Figure 6.6. On December 2, 1933, after a horrific day of route-finding through crevasses, the party camped on Scott Glacier abreast of Mount Russell, pictured here in the foreground, with its fractured umber face of pure granite. To the rear the ridgeline rises to the blocky western wall of Mount Blackburn, the most dominating massif on the eastern side of Scott Glacier. Prominently displayed midway up Blackburn's face is a thick succession of Beacon sandstone sitting on the old erosion surface on the granite.

they climbed, the wind direction shifted from south to east. Again the party was faced with high longitudinal pressure ridges on the glacier and numerous crevasses transverse to its flow. Travel was tedious, but most of the crevasses had bridges filled to the bottom. By 1:00 P.M. the party had traveled eleven miles. The wind was becoming so unbearable that the men decided to pitch camp, feeling as they did that they had come through the worst of the crevasses. That day they had stopped on ice north of the confluence of Scott and Robison Glaciers (Fig. 6.7; see Fig. 6.2).

The evening's scheduled radio contact brought a long message from Little America advising them to quit the place where they were and to proceed to the vicinity of 85° S, 115° E, a distance of about two hundred miles as the skua flies, "to get some information on the connection between East and West Antarctica." In the time-honored tradition of the independent field party, Paine failed to acknowledge receipt of message.

Here the men were, within striking distance of the sedimentary strata, each day bringing its own fresh wonders. Visible for the first time this afternoon on the east side of Scott Glacier was a lofty ridge that climbed up from the granite spurs at glacier level. Flat-topped and massive, its crest was capped by a brown sill of dolerite above a thick section of sandstone and shale. To reach these rocks, perched as they were high above the glacier, would require at best a long and steep approach over the granites. Tonight to the southwest, however, small, dome-shaped mountains barely marring the horizon appeared from this vantage to arise from the surface of the glacier. There would be the place for sedimentary rocks to be had. No chance the party was going to abandon its course to check on the connection between East and West Antarctica. From that last, low roll in the south, the men were destined to discover the southernmost rocks on the planet.

The following morning, December 4, the southeasterly winds were again bearing down at twenty to thirty knots and the temperature had slipped to minus 1° F. Except for ground drift the air was clear. The party lay in for the morning, but restless to be moving, the men turned out at noon into the teeth of the wind. In all they made thirteen miles that day, over large pressure swells along the northwest portion of the La Gorce Mountains, stopping to make camp near the end of the long spur running north from Mount Gjersten (see Fig. 6.7). On clear ice nestled in among boulders of a scant moraine, the

Figure 6.7. (overleaf) On December 3 the party camped at the confluence of the Scott and Robison Glaciers, seen in the small curl of flowlines in the glacier at the left rear. Mount Gardiner is the dark shoulder on the right (west) side of the glacier. Dominating the left half of the horizon are the La Gorce Mountains. The following afternoon, under heavy winds, the party moved thirteen miles farther along, camping near the end of the spur leading toward the viewer from the highest central peak in the La Gorce Mountains. The next day, uncertain of a passage through the rolling icefalls that stand out between the La Gorce Mountains and the two nunataks in midstream, Blackburn and Paine went out on skis to scout a route. They found a passage that cut south-southwest across the glacier, ending at a small, tabular outcrop, Mount Wilber, seen to the right of the summit of Mount Gardiner. Two days later, December 6, the party was camped at the base of Mount Wilber, and from there moved over to Mount Weaver, whose eastern spur shows slightly at the edge of the image. Mount Howe, the southernmost rock on earth, is the faint, tabular outcropping at the horizon, framed by the two small nunataks in midstream.

dogs and men clung to their spot. Blackburn logged the next day that "Wind blew a half gale on last nite threatening to blow away the tents. But they held being anchored and held down by rox."

The next morning the wind was nearly as strong, so rather than break camp Blackburn and Paine hiked southwest to check the route that lay ahead. Having rounded the end of the spur where they were camped, the men found themselves in a calm area where apparently the wind seldom blew. Scrambling over ice-cored moraine and pressure swells to gain some elevation, they were able to see that a clear way lay to the southwest, directly toward a dome-shaped mountain capped by sedimentary strata, and that not far ahead, the blue ice, so persistent up until now, gave way to snow-covered glacier (see Fig. 6.2).

The men returned to break camp, and by 3:00 P.M. were driving south again. At 5:40 the party stopped to prepare for a scheduled radio check. Again Little America suggested that the party abandon its course and head eastward, to which Paine "replied in the negative." He fumed in his diary, "It's absurd—75 miles up, almost to the plateau. Our goal in sight, nine days in the field gone + he wants us to give it all up—bunk directing."

The following day was an excellent run of nineteen miles, bringing the party to the northern end of Mount Wilber, a ragged outcropping of granite with wild drifts clinging to the upper reaches of its eastern face (Fig. 6.8). A few miles to the west, beyond a broad icefall, rose Mount Weaver (Fig. 6.9). Its base was granite, but above the unconformity fully 90 percent of the mountain was sedimentary strata, layer upon layer of tan and green and brown, rising finally to a summit capped by a sill of umber dolerite. Compared with the massifs that the men had passed in the middle reaches of Scott Glacier, Mount Weaver was diminutive and plain. Its summit was nearly flat and its overall appearance blocky. From bottom to top the smooth, north face measured about two thousand feet, with a prominent spur leading down from the summit toward the northeast.

Figure 6.8. Mount Wilber stands as one of the first obstructions to the draft of katabatic winds that funnel into upper Scott Glacier. At its eastern end, wind running up the backside is compressed and wind that rounds the corner at a lower elevation is draw by the Venturi effect vertically up the rocky face, producing wild drifts like horsetails that end at the upper edge. Blackburn's party camped under this face for one night before moving over to Mount Weaver.

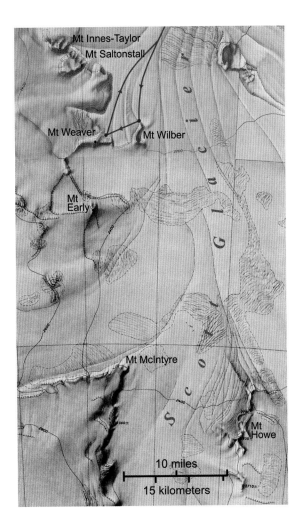

Figure 6.9. Shaded-relief map of the area around Mount Weaver, showing Mount Howe, the planet's southern-most outcropping of rock.

Though it would take its due, this was not a mountain to challenge the body and the will, not one to which the spirit rose for grandeur. This was more a mountain of the mind. Here was history, succession, order, the possibility of life enshrined, preserved as fossils through the eons. And if fossils, then the notion of time, and where this mountain fit in the grand history of mountains on Earth.

The next morning with wind pouring down at thirty knots and the sky gray and overcast, Russell hiked over to the moraine at the foot of Mount Weaver to see whether he could find a sheltered campsite. He returned shortly after noon with a lump of coal, one of many, he reported, that littered the ice, and he revealed that a suitable campsite existed. Anxiously, everyone loaded up and moved on in to the base of the mountain, laying camp one hundred feet from the edge of the moraine.

Fifty-three days, 727 miles out from Little America, this would be the farthest camp. With six days of food at hand before the need to start back, the men had a little time to give the weather a chance to settle. For a shot at the summit of Mount Weaver and a collection of the complete stratigraphic section, the party needed the best possible conditions.

During the first afternoon the men ambled through the moraine, collecting. Here was a varied sampling of the rocks cropping out in the bedrock above, a potpourri du Mount Weaver, evenly mixed and thinly spread over gray ice. The main ingredients were sedimentary rocks, sandstone, shale, limestone, coal. For flavor, boulders of dolerite and granite were scattered throughout. But the most delectable morsels of all were bits of fossil plants, impressions in dark shales and carbon films in some of the sandstones. The farther the men walked, the more they found. Most was pretty scrappy material, but here and there part of a rare leaf was preserved. The moraine was huge and deserving of a thorough look.

In the evening the party celebrated its farthest south with a feast of goodies carried all the way for just this night. Blackburn recorded the menu: "Beef hamburger, tea with lemon, hoosh with pemmican & hamburger, chocolate ice cream, bacon, biscuit & a bottle of Sherry (also a little alcohol)."

Blackburn: "Dec 8/34. 8:45 PM. Spent the day collecting on north side of Mt. Weaver, mostly on moraine. Will this infernal S.E. wind ever quit?"

The following day the sun broke through, but the wind remained a miserable twenty-five knots. The party lay in for most of the day writing letters to family and friends, but turned out for some group photos with the flags flying in the afternoon. In the evening Blackburn again was out collecting fossils. The wind had not changed. When he thought back over the Scott Glacier traverse, it seemed to him that precious few of the days had been calm at all. In fact, perhaps the calms had been a matter of place, and the wind, after all, was a constant. The steadiness of the wind at Mount Weaver was unsettling. Seldom if ever were there gusts, just an unrelenting draft from the plateau, down the rocky face of the mountain and over its flanking icefalls. Perhaps here the wind never did stop. Little time remained to find out. So enough of this waiting. Tomorrow Mount Weaver would be climbed . . . and the wind be damned.

The next morning, December 10, 1933, Blackburn, Paine, and Russell were ready. The temperature was 1° F, the skies were clear, the wind blew at twenty to thirty knots. Dressed in an extra layer of everything, and toting a light tent, Sterno cans, medical kit, compass, barometer, thermometer, climbing rope, ice axes, and several hundred sample bags, the party trudged out to conquer the mountain. The northeast ridgeline offered the best route, although it was dreadfully exposed to the wind. At times the men were blown off their feet, and had to cling to the rocks to avoid being tossed down the face below the ridge crest. Worse still, the intensity of the wind increased as the men climbed, peaking about halfway up the mountain.

Beyond that, mercifully, the wind dropped off, even though the men could see that drift was still swirling around their tent far below. At the summit the air was almost calm. For the first time since the ice shelf the men had risen above the layer of inversion winds that pour off the plateau and down the tributaries into Scott Glacier. Even though the temperature was minus 14° F, the cold did not penetrate. The peacefulness gained through the ascent intensified the vista. On the far side of the summit the men glimpsed for the first time over the edge of the plateau, out onto the vast whiteness that spread to the southern horizon.

THE WIND

In the static, lifeless landscape of the deep field, the wind is the only animate force. It is movement and sound, alternately relentless and fickle. When it stops and the sun beams down from a cloudless sky, you can strip to bare skin and immediately feel the warmth. But let one puff of breeze disturb the thin layer of radiant air, and shivers will well up. When the wind picks up, it buffets the parka and bites at fingertips, ear lobes, and nose. In its full fury the wind has flattened tents and thrown men from the decks of ships. At these times it is an awesome, fearsome force.

During the 1980–1981 field season I was camped between Mount Mooney and the La Gorce Mountains a few miles from our put-in site on Robison Glacier. For the better part of the two weeks we spent at that camp, frigid, katabatic winds poured over us from the polar plateau to the southeast. With wind speeds generally around ten to fifteen knots and temperatures about minus 10° F, our days mapping the outcrops in the surrounding area were seldom comfortable, especially along ridgelines where the wind compressed and accelerated.

The La Gorce Mountains at the edge of the polar plateau are a first obstruction to katabatic winds that originate deep in the interior of the East Antarctic Ice Sheet. The flat top of these mountains slips smoothly from beneath the ice sheet and rises to the northwest to a dramatic escarpment that drops steeply more than three thousand feet and splays into two major ridge systems (see Fig. 6.2). When the katabatic winds meet the southeast or back side of the La Gorce

Figure S.11. A cloud layer shoots out from the escarpment lip of the La Gorce Mountains. Beneath it, a plume of snow traces the dense, frigid, katabatic wind as it leaves the precipice and plunges into the valley behind the intervening ridgeline.

Mountains, they split into three streams: two follow the descending glaciers on either side of the mountains, and a central stream shoots across the flat summit and plunges off the lip of the escarpment.

From our camp on the glacier we looked up to the escarpment. On days when the wind was light and there was only a trickle of granular, blowing snow around base camp, we could see a churning plume of wind and snow plummeting from the escarpment lip high above (Fig. S.11). We would watch it and imagine that somewhere back from the edge beyond where we could see there was a valve that tapped the source of all winds, screaming as it released its jetted fury.

Figure S.12. Blowing snow drifts over base camp in the La Gorce Mountains, December 1980. The northern escarpment of the La Gorce Mountains is visible to the left rear. The Scott tent in the middle of the image served as the cook tent for our four-man party, while we each had an individual mountain tent for sleeping (not in view). To the left of the Scott tent are two Nansen sleds with triwall cardboard boxes containing food. Two snowmobiles, one covered, the other with its windshield exposed, sit behind a third Nansen sled. On the right, a member of the team is taking a shovel out of a wooden box mounted on a Nansen sled to dig an entrance to his sleeping tent. Behind that is another Nansen sled with a shock of bamboo poles and flags for marking trails.

During one three-day period, the wind speed rose to forty knots around camp, and we were forced to hunker in our tents, enveloped in blowing snow. It is during storms like these that I have learned to love the Scott tents. When planted properly, these four-sided pyramids will bear the fiercest gale, their double walls flapping loudly as they keep out the force of the wind. During this storm, tumultuous clouds ripped through the scene, opening periodically to reveal the escarpment lip beyond the adjacent ridge (Fig. S.12).

A storm such as this can move in quickly, so we always have to be cautious if working far from camp, and watch that the weather doesn't turn. To be caught out can truly be a matter of life and death. But back at camp with the warmth of the cookstove at hand we can feel secure and even cozy. Then it is good to go out into the blast, not to confront the wind but to feel its pressure, to lean the body into it, to find the angle of balance, to sense the vagaries in the flow, to feel the cold, to listen to the voices wheezing and whistling around every obstacle in camp—tents, boxes, bamboo poles.

Out beyond the noise of camp, we hear only the soft shoosh of blowing snow streaming through the sastrugi. We look up to a blue sky and down into the miasma of snow and wind at ground level, opaque beyond one hundred yards. We are walking at the dynamic interface of atmosphere and solid earth; wind pants flap, and we squint with one eye peering down the tunnel of the hood, balancing between the cross gust and the pitch of our strides. Noses drip and fingers begin to ache from the cold. The wind is right there with us: we slip on through its stream.

Where the ice began to spill off the plateau, broad swaths of crevasses foreshortened to narrow bands were glistening in gray and white, like the scales on schools of so many fish. Just at the horizon the empty blue sky converted to white, slightly less brilliant than the ice sheet beneath. About five miles directly south of Mount Weaver was a faceted peak, symmetrical and sharp-topped, covered to the summit on its northwest face by a graceful drift (Fig. 6.10). Another ten miles south of that, the headreaches of Scott Glacier merged with the plateau between a low ridge mainly covered in ice and a more formidable edifice of rock standing like a black wharf in the sea of white (Fig. 6.11).

All the southern continents have a land's-end—that last, narrow cape before the Southern Ocean, for land dwellers a place where souls depart this earthly realm and seafarers the gateway to beyond. Mount Howe is the land's-end of Earth, interior to Antarctica, the last promontory from which one can gaze south across the frozen ocean, out toward the still turning beyond the arced horizon. It creates the first turbulence in a flux of air begun at the outer edge of the troposphere. Surely this is a place of spirits—that is, if any inhabit this frozen region at all (Fig. 6.12).

For 360° the Earth radiated from the summit of Mount Weaver. The climbers were

Figure 6.10. The faceted cone of Mount Early stands at the edge of the polar plateau. In this view of its eastern face, the drift that Blackburn recorded from the summit of Mount Weaver runs off the right side of the image. Unknown to Blackburn at that time was that Mount Early is a volcanic cone erupted beneath the ice sheet, the world's southernmost volcano.

Figure 6.11. Mount Howe stands beyond the summit of Mount Early like a wharf at the edge of the great icy sea. This is a view at about the same angle, but closer by about ten miles, to that witnessed by Blackburn's party from the summit of Mount Weaver.

filled with the euphoria of such a place—the release from the sustained struggle, the attainment, and the vista surpassing the visual.

Paine later wrote of the view:

The mountains which fringe the valley of Thorne [*sic*] Glacier stood in serried ranks, like matched soldiers, the shorter in front with the bigger and huskier well back of the front rank. The main escarpment of the plateau stood higher than the rest. It seemed to form a background to the soaring granitic peaks nearer the main glacier, making

them lean and tall against its own massiveness. Each of us felt a sense of elation as he took in that panorama of mountains, glistening snow domes, glaciers and skies. There was a serenity and peace on the land. . . . Soft blues and greens merged into the dazzling whiteness of snow above, and purple lines told of fissures and grottoes. . . . Sweeping around the northern base of our mountain from the east was a tributary glacier. From our lofty perch the broken waves of ice and crevasses appeared like ripples on the smooth waters of a slowly moving river.

At the summit of Mount Weaver the men built a cairn, placing their names in a cocoa tin. Beside it they lunched on pemmican and chocolate, lingered a while longer. But the sweet music that had lured them into the unknown had stopped. On the way back they would have to pay the piper—in the tender of geology with triangulations and samples of rock.

Section measurement is a tedious business, determining the thickness of a unit of stratum, describing its characteristics, collecting a representative sample, then doing it all again, and again, until the section is completed. Under Antarctic conditions teamwork is essential for efficient collection of such data. Blackburn measured thicknesses, dictated descriptions that Paine wrote down, and chose the samples that Russell labeled

Figure 6.12. Earth's land's end, a stark landscape of ice, rock, wind, and the mind. Mount Howe marks the horizon to the left of the image. Mount Wilber is the black rectangle directly above the figure. To the right of Mount Wilber is a broad icefall. To the right of that, Mount Weaver silhouettes its eastern ridgeline, climbed on December 10, 1933, by Blackburn, Paine, and Russell to the pointed summit.

Earth's Land's End

and bagged. The summit was capped by 130 feet of dolerite, but a short distance below was a shale layer chockful of plant remains. Below this the section was an alternation of sandstone, shale, and coal. The plant fossils were later identified as *Glossopteris,* a seed-fern of Permian age that was known in other Southern Hemisphere continents and in India, and was first identified in Antarctica among the rocks collected along Beardmore Glacier by the men of Scott's party, and carried to their death.

Blackburn handled rocks so much on the way down that his fingers became frost-bitten through his mitts. In all he distinguished fifty separate units over more than one thousand feet of section, and collected sixty-seven samples, although the base of the section above the eroded granite was covered by snow at the end of the spur. The load of rocks helped the men keep their footing on the windy ridgeline, but they were so loaded at the bottom that they could barely walk back to camp.

The following day the men organized their gear and rested from the rigors of their climb. Paine: "Spent day bragging + marking specimens. Temperature continues zero + below with incessant 20–30 mile wind. Glad to leave this spot. Camp Poulter—peace we hope tomorrow night."

On December 12 they broke camp and started the long trek back. In a favorable position several miles along the trail they made their first stop to take triangulations that would be the basis for the map of the Scott Glacier area (Station 2) (Fig. 6.13). With a trailing wind and a gentle downgrade on the glacier, the party covered 21½ miles the first day. The wind had blown at about twenty knots when they started, but it lessened to barely a breeze by the time the men pitched camp abreast of Mount Grier (Station 3).

The following day was calm, and both men and dogs lounged in the warmth of the sun. Blackburn took another round of transit readings and then a round of photos. Paine and Russell cut each other's hair. Everyone was feeling good. From there they descended Scott Glacier with relative ease, stopping every five to ten miles to triangulate.

On December 14 the party pulled down to its old camp on the moraine at the confluence of the Scott and Robison Glaciers, where the men shot a round of bearings (Station 4), then moved on to Station 5 and camped in a broad area of ice where Van Reeth Glacier joins the confluence of the Scott and Robison. The next day Paine took a prime vertical shot at about 3:45 A.M. and another at noon, and Blackburn triangulated the surrounding peaks for about three hours in the morning. After noon the party left the camp under a bitter southeast wind, but within about an hour a draft of warmer air moved up the glacier. The solstice was approaching, and weather patterns had begun to push south from the ice shelf, bringing warmer temperatures, stratus clouds, and surface fogs that intermittently crept up Scott Glacier (Fig. 6.14).

That evening the party camped abreast of the blocky massif that Blackburn named Mount Jessie O'Keefe, with "mountain after mountain and range upon range in view up and down the valley." Byrd later renamed the mountain Mount Blackburn, one of numerous changes to the names that Blackburn and the others in the party had chosen for the features they discovered (see Fig. 6.13). Each of the party members named a prominent peak after a woman. Jessie O'Keefe was the woman Blackburn would marry. Katherine Paine was Stuart's mother, but Byrd shortened the name to Mount Paine, after Stuart himself. Mount Jane Wyatt, named for a young actress from New York City who would

Figure 6.13. (opposite) The sketch map published by Blackburn in 1937 in the *Geographical Review* shows the descent route on Thorne [*sic*] Glacier with his numbered triangulation stations clearly labeled. According to the caption to the figure, "Because of the great wealth of material in the form of photographs and triangulation data furnished by the geological party and the fact that some features have been identified on the aerial photographs taken on the South Pole Flight of November, 1929, the map has not yet been completed. Pending its completion it should be noted that names are not final." In fact, the expected map was never completed, and a number of the names on Blackburn's sketch map were subsequently changed by Byrd and the Antarctic Names Committee. Reproduced by permission of the American Geographical Society.

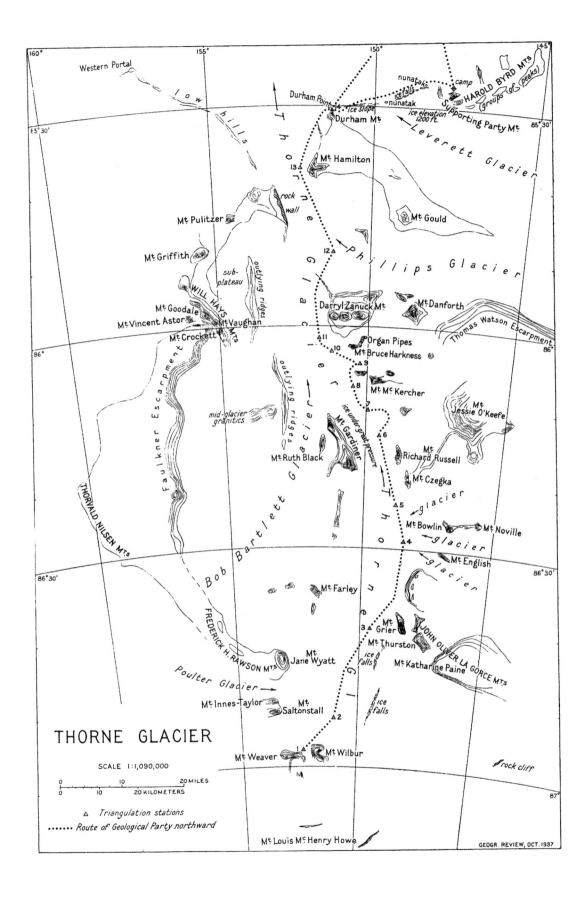

THORNE GLACIER

SCALE 1:1,090,000

△ Triangulation stations
⋯⋯ Route of Geological Party northward

GEOGR. REVIEW, OCT. 1937

Figure 6.14. Intermittent stratus clouds billow over the summit of Grizzly Peak in the foreground and over the foothills of the Hays Mountains on the far side of Scott Glacier. In the distance the shadowed face of the Faulkner Escarpment demarcates the eastern margin of the Nilsen Plateau.

go on to have a long and distinguished career in movies and television, was shortened to Mount Wyatt, with the citation in the official gazetteer of names reading, "for Jane Wyatt, a friend of Richard S. Russell, Jr., a member of the party."

December 16 was "a day of mist and rolling clouds," with intermittent, light snow. The peaks across the glacier were mostly obscured, but they appeared enough for Blackburn to triangulate at Stations 7 and 8. On December 17 the party reached the depot at Mount Harkness after passing through about a mile of heavy crevasses. From there they pulled about three miles up to the east into the center of the amphitheater of peaks that they had spotted on their southerly traverse (Fig. 6.15; see Fig. 6.5). The sastrugi all but disappeared, indicating the wind seldom reached into the sanctuary. Paine called it the "most attractive of all spots the world over." The peaks that rimmed the amphitheater evoked the notion of a Gothic cathedral, the men agreed. Both Blackburn and Paine mention the resemblance in their journals. Blackburn: "This gathering basin is walled by towering peaks which, due to the apparently vertical jointing of the brownish granitic rock, have assumed huge towering spires buttressed after the manner of gothic cathedrals." Paine: "To our north and south are granitic peaks with nearly vertical cleavage, making them stand up like walls or pinnacles like organ pipes. To N.E. five miles away is a row of these gothic pinnacles perhaps 800 feet high above us, standing like immense gothic towers all in a row" (Fig. 6.16).

Figure 6.15. Shaded-relief map of the Gothic Mountains, showing the location of the camp in this "most attractive of spots the world over." The camp was erroneously located on Blackburn's sketch map (Station 9) to the south of Mount Bruce Harkness and the Organ Pipes. On that map the camp should have been spotted between the Organ Pipes and Mount Darryl Zanuck (compare with Fig. 6.13).

The dogs were in bad condition and enjoyed lying about basking in the sun. After dinner, while Paine and Russell wrote in their diaries and lounged, Blackburn skied over into the small valley to the south of the Organ Pipe Peaks. As if constructed as a chapel to the larger sanctuary, the vale led back to a faceted peak of dramatic continence, glimpsed by the party on its way up to camp. As Blackburn glided along on his skis, the only sound was the swooshing of the boards and the shushing of his breath. The sun shone low from the western sky, illuminating the peaks in a glow of burnished gold. The central spire cast its aura on the spot (Fig. 6.17). Blackburn stopped. His breathing lightened. Stillness descended.

Figure 6.16. The Organ Pipe Peaks stand in ordered splendor in the sanctuary of the Gothic Mountains.

Figure 6.17. The Spectre casts its aura, awaiting awestruck seekers in its icy chapel.

The only sound was the soft scrunch of the pulse in his ears. With its arched vertical pitch of two thousand feet, The Spectre towered above and slightly forward of the lesser deities attendant at its flanks. Fleecy clouds rolled up and over the backside of the peaks, adding motion to the silence. After paying proper homage, Blackburn unpacked the Zeiss, shot the "fine photo effects" for later revelation to the world, "and slid on back to camp on the long boards."

ASCENT OF THE SPECTRE

I remember the first time I saw a photo of the Organ Pipe Peaks. It was 1970, and I was a graduate student due to leave on my first expedition to Antarctica. I did not believe that any grouping of summits could be so dramatic, beautiful, and perfect. They were a fantasy of mountains rendered with bold and simple strokes, faceted grandeur in black and white. I dreamed of traveling to those peaks, bowing down before their central spire—The Spectre—and sampling a piece of the rock. Alas, the Organ Pipe Peaks remained beyond my reach that season.

Ten years later on my fifth Antarctic expedition, I found myself camped on Sanctuary Glacier, in the shadow of the Organ Pipe Peaks (see Fig. 6.15). My work that season had begun with geological mapping of the La Gorce Mountains (see Fig. 6.2) and ended with a collecting traverse down the east side of Scott Glacier, essentially retracing Blackburn's route. Now I was one day away from attempting an ascent of The Spectre (see Fig. 6.17)—the splendid spire that had awed the Blackburn party.

Being a klutz with ropes myself, I have always had someone in the party who is experienced with roped climbing in case we needed it. This time, the field assistant/mountaineer for the party was my brother Mugs, who had the year before made his first big mark in the climbing world with the first ascent of the Emperor Face on Mount Robson in British Columbia. From the beginning of the field season, Mugs and I had joked about climbing The Spectre. We figured he would do all the leading, and if necessary would winch me, the older brother, up on the rope. Now that we were camped in the shadow of The Spectre, looking up its backside, the climb was no longer a joke. It was real, sheer, and daunting. Mugs studied the fractured upper wall of the spire, and, although he couldn't see a clear route, said "we'll just wander around on the face and see where it leads."

I understood Mugs's nonchalance and trusted him completely. I also trusted myself. I must admit, however, that I didn't sleep well the night before the climb. What would it be like? Would the rough passages be vertical or overhung? Would I be in way over my head? I hadn't had such a case of butterflies since before wrestling matches in high school.

After a big breakfast Mugs and I snowmobiled over to the foot of The Spectre (Fig. S.13). We carried a minimal rack of climbing gear: a half-dozen carabineers, several slings, and four pitons to secure the rope. The first half of the ascent was a straightforward climb up a steep (50°) snow chute to a shoulder on the right skyline, with Mugs kicking in all the footsteps and I following in his prints. At the shoulder we pulled out the rope, and while I belayed, Mugs began working his way across and up the face, which in this stretch was pretty much vertical. When he reached secure spots, Mugs would set the belay for me, and I would follow up his path. There were good-sized cracks in the rock that gave plenty of handholds

Figure S.13. North face of The Spectre, the central spire of the Organ Pipe Peaks, showing our ascent route.

and places to rest, so I mostly managed to climb with no problem. The most difficult passage of the climb—the crux—occurred at a place where there was a slight overhang. The only handholds were high above my head, but there was nowhere to place a toe if I pulled myself up. I thrashed around some as Mugs laughed and tightened the rope. But then I found a bulge on the rock out to my left side that could be grasped between my knees. From there I could reach the next handhold, and we were both past the touchiest part of the day.

After about two hundred feet of roped climbing (two pitches), we made it past the steepest stretch, and came out onto a rock face with a slope angle closer to 60° than to vertical and with lots of snow-filled cracks that made planting steps easy. Here Mugs packed the rope, and we continued upward across the face (Fig. S.14). We had started the day in full sun, climbing in shirtsleeves with our parkas packed, but as the sun circled its way to the south, we slipped into shade and the chill that it brings. Rather than take off our packs on the steep terrain, we decided not to pull out our parkas, and pushed on to the summit. A small cornice of soft snow maybe eight feet high was the last barrier to the top. Mugs chopped and kicked his way through and over it, and we emerged into sunlight on the flat of the summit. In all directions splendid peaks reached for the heavens, piercing the undulating mantle of white and blue. No sound stirred the silence. We took a round of photos and then had some lunch.

You could say that we were pleased with ourselves. I can also say that we brothers never felt closer. Each of us knew that he wouldn't be at this spot were it not for the other. I had provided the opportunity and Mugs the expertise. What I recall most was agreeing with Mugs that our parents would be more than doubly proud. We lingered a bit longer and finally descended; I rappelled most of the

Ascent of The Spectre

Figure S.14. Here I am free-climbing the upper portion of The Spectre. Photo by Mugs Stump.

distance down to the shoulder, and Mugs mostly downclimbed after me, stripping the hardware from the belay points. At the shoulder, we figured a glissade down the snow chute would be the fastest way back to the base of the mountain, so we sat back on our heels, set the points of our ice axes in the snow for braking, and slid all the way down to our snowmobile. My dream a decade earlier of bowing before The Spectre had been exceeded. What a grand and memorable day in the mountains it had been!

December 18 dawned a "beautifully clear day among mountains." A bank of fog lay six hundred to seven hundred feet below on Scott Glacier, and climbed up the valleys between the spurs of the escarpment on the western side. The rations for both men and dogs were more than ample to reach their next depot at "Mount Base," so the party lingered another day.

Here Blackburn assessed the geology he had encountered during the expedition. The entire width of the range appeared to be underlain by igneous rocks (Fig. 6.18). Most were reddish-brown, some were darker, but all appeared to be plutonic. They indicated that in this region a massive tract of crust had melted, forming a composite body of granite plutons, a *batholith*. The scale of the body was comparable to other great granite masses on Earth such as the Sierra Nevada batholith of California or the Patagonian batholith of the southern Andes. (A batholith is a body of intrusive igneous rock of large, regional dimension.) The small occurrences of gneiss and schist to the north at Mount Hamilton

Figure 6.18. This is the view of the mountains that Blackburn saw while sitting in camp on December 18, 1933, looking west out the entrance to Sanctuary Glacier, across Scott Glacier to the other side. Every exposure of bedrock in sight was the same blocky reddish-brown, characteristic of granite he had seen throughout the region.

and Supporting Party Mountain and to the south in the La Gorce Mountains were the country rocks into which the granites had intruded, perhaps from which they had been melted as well. The mountains formed during this episode of granite generation were subsequently eroded to a surface of low relief upon which the overlying sedimentary strata had been deposited. Blackburn recorded that the erosion surface beneath the sediments now lay at about the same elevation on both sides of Scott Glacier, indicating to him that the Transantarctic Mountains had been uplifted en masse in this region with the glacier subsequently cutting down through them.

In the afternoon, Paine and Russell skied over to the vale beneath the Organ Pipe Peaks. Equally awestruck, they lingered. At the head of the alcove at the foot of Altar Peak, Russell discovered lichens in a crevice in the rock, at that time the southernmost known life. On one of the steeper undulations in the valley, the slope ran out for about a quarter of a mile. The intrepid pair made three runs at "a speed of 40–50 miles an hour," after which they took picturesque shots of the mountains, including some with Grape Nuts and products of other sponsors as foreground interest. The fog undulated slowly up the glacier and by 3:00 P.M. had enveloped camp. Paine and Russell were caught out slightly higher up, but the fog hovered not much beyond camp, so they made it back without a problem.

Encircled by the Gothic Mountains, contemplating the isolation and the solitude,

Figure 6.19. Reaching toward a certain perfection, a nameless spire of the Organ Pipe Peaks beckons at the portal to the Sanctuary.

free of any fear of place, Paine wrote in his diary, "We all feel marvelously self-sufficient, contented + I believe truly happy. There is no peace like the peace brought by complete separation from other people. We are nature worshipers, perhaps pantheists" (Fig. 6.19).

On December 19 the party bid farewell to the sanctuary, and eased down onto the margin of Scott Glacier. Stopping twice for triangulations, they made eighteen miles before stopping about three miles beyond their "Granite camp" of the outward traverse. On the moraine at the northern tip of Mount Zanuck, the party collected specimens. They also checked the array of flags that they had laid out on their ascent, but could detect no movement. Paine made a scheduled radio check. The next morning the men awoke to find four skuas visiting them. After shooting a round of points, they eased down to Mount Hamilton, where, in addition to rock specimens, they collected numerous lichens. On the 21st the party was back at Durham Point. The men climbed the peak behind it in the evening, with a spectacular view of the lower reaches of the range along the ice shelf.

On December 22 Blackburn logged, "Left Durham Pt. about 10:30 A.M. and came over to the Mt. Aviation Base. Hence we are back once more on the flagged trail ready for the return. Collected lichen at Durham Pt., took a round of angles to prominent peaks, shot the sun for Lat., packed up; and thus we quit the Mts."

They completed the journey across the ice shelf in record time—527 miles in sixteen days. The party traveled mostly at night with the sun at their backs and a harder surface underfoot. A torn tent was strung on bamboo poles and used for sails on the sleds, speeding the party along. Paine: "With our sails making us like Arabian fishing smacks we fairly bobbed along. . . . Each flag stands as a buoy, the beacons + depots as harbors, lightships or refueling ships, the sastrugi are the waves + it is truly remarkable how similar in sound is the plunge + the wild cavortings of the lead sled and that of a small boat in a choppy sea."

Along the trail of depots, food amassed faster than the men and dogs could consume it. They were putting on weight and sleeping late. On January 10 the Condor touched down, took a sounding, and relieved the party of 500 pounds. The following day, after a run of forty-five miles, the party hauled into Little America at 7:00 A.M., healthy as ever, fitter by far, and beaming with 450 pounds of rock on their sleds.

Even more than the first expedition, Byrd's Second Antarctic Expedition bridged the heroic and the modern eras of Antarctic exploration. Mechanical ground transport proved itself for the first time, and radio communication was by voice rather than Morse code. Air transport revealed vast areas of previously unknown West Antarctica and facilitated the geophysical sounding of the ice sheet. Aerial reconnaissance also established that the Horlick Mountains lay 170 miles farther than the previously observed limit of the Transantarctic Mountains. The deepest penetration into the continent by Blackburn's party was, however, still accomplished using dogs. The bulk of the geographical discoveries of BAE II was the extended mapping of Marie Byrd Land. Nevertheless, the traverse of Scott Glacier by the geological party stands as a triumph of polar exploration, with the discovery of the magnificent Gothic Mountains and Mount Howe, the southernmost rock on the planet, the recognition of the tremendous extent of the Queen Maud batholith, and the large quantity of rock samples. Laboring under severe katabatic winds and crossing crevasse fields as daunting as any encountered by their predecessors, Blackburn, Paine, and Russell accomplished their journey with a competence and grace that has never been equaled.

7 To the IGY and Beyond
Filling in the Spaces

With the conclusion of Byrd's Second Antarctic Expedition, the full extent of the Transantarctic Mountains out to the Horlick Mountains had been discovered. To be sure, broad portions of the map remained blank on the plateau side of the mountains, a gap existed between Beardmore and Liv Glaciers, and the distal termination was vague and uncertain, but explorers had mapped the Transantarctic Mountains for fifteen hundred miles from North Cape to the Horlick Mountains, and had made crossings to the plateau at six locations: David Glacier, Ferrar Glacier, Beardmore Glacier, Liv Glacier, Axel Heiberg Glacier, and Scott Glacier. Subsequent comers to this once virgin land would largely be filling in the gaps.

The exploration of Antarctica during the first third of the twentieth century was primarily funded from such private sources as learned societies, wealthy donors, or contributions from the public at large. Nevertheless, national interests were also at play on many levels. Gould's claim of Marie Byrd Land (lodged in the cairn at Supporting Party Mountain) was only one of many left by private expeditions in the name of the United States of America. Claims and flags were dropped from the air by Byrd, as well as Hubert Wilkens and Lincoln Ellsworth, during their series of airborne expeditions to the Antarctic Peninsula and West Antarctica in the late 1920s and 1930s. While the government did not publicly press these claims, behind closed doors it grappled with the issue. Great Britain, New Zealand, France, and Australia made formal claims to portions of Antarctica between 1908 and 1933 based largely on discovery. Norway followed suit in 1939, in response to German activity in an area that their whalers had hunted. In positioning itself to assert its own claims, the United States adopted a position that modern claims should follow "constructive occupation," including the maintenance of permanent bases and the carrying out of administrative acts.

By the end of the 1930s the U.S. government moved to further its interests in Antarctica by establishing the United States Antarctic Service Expedition. The service was a civilian organization administered by four cabinet agencies. Scientists came from the civilian sector, but the U.S. Navy conducted the expedition, with Admiral Byrd in command. Although funding was mainly from congressional appropriations, Byrd offered a variety of personal equipment from his previous expeditions, and substantial contributions also came from private donors, such as Charles R. Walgreen of the drugstore chain and William Horlick of malted milk fame.

The expedition, numbering 125 men, two ships, and four aircraft, was charged by the Congress and President Roosevelt to "investigate and survey the natural resources of the land and sea areas of the Antarctic regions." Two bases were established with the intent of their being occupied more or less continuously, one at Stonington Island on the west side of the Antarctic Peninsula (East Base), the other at Little America (West Base). From these two locations numerous flights and overland traverses filled in much of the unknown territory of coastal West Antarctica, with the crews and parties leaving claims in canisters followed by copies duly filed at the State Department.

One flight was planned from West Base to the Transantarctic Mountains with the intent of closing the gap between Beardmore and Liv Glaciers. On February 29, 1940, the Condor flew southwest across the Ross Ice Shelf bound for the eastern portal of Beardmore Glacier. On board were Paul Siple, the leader of West Base, and F. Alton Wade, geologist and chief scientist for the expedition. Siple had been the official Boy Scout who accompanied Byrd's First Antarctic Expedition and had returned on BAE II as a biologist and dog driver. By 1939 he had received a Ph.D. in geography studying the effects of cold on men in Antarctica. Wade had been a dog driver on BAE II, and subsequently received his Ph.D. in geology.

When the plane was about 150 miles out, the mountains began to appear at the horizon. Directly to the south, one prominent massif stood out above all others, remaining in view for the duration of the flight along the mountains. As the plane approached the mouth of Beardmore Glacier, the men recognized Mount Hope and The Cloudmaker among others mapped and photographed by Shackleton's and Scott's expeditions (see Figs. 4.10, 4.12). Climbing to an altitude of ninety-five hundred feet, the plane circled for a round of photographs, then turned southeast and flew along the mountain front.

The foothills maintained a low, relatively uniform elevation back to an escarpment that rose abruptly to heights well above the altitude of the plane. As the plane approached the massif that the men had spotted from afar, an outlet glacier appeared abruptly, cutting a deep swath straight through the mountains. To the east beyond this glacier, the landmark summit, later named Mount Wade, commanded an array of diverging ridgelines that filled the foreground (Fig. 7.1).

The plane circled for another round of photos, then flew into the drainage of Liv Glacier. The gap between Shackleton's and Byrd's sightings had been closed with the discovery of a major outlet glacier, thereafter named the Shackleton. Airborne for eleven hours, the Condor touched down safely at West Base with fifteen minutes of fuel left in its tanks.

Figure 7.1. First seen in December 1939 during a flight to close the gap between Beardmore and Liv Glaciers, 13,398-foot Mount Wade towers over its domain. Shackleton Glacier drains from the rear to the right of the image, behind Mount Wade. Mount Munsen, used as a survey station by the Topo East party in 1963, is the dark-faced peak in front of and to the right of the summit of Mount Wade.

Although it was the intention of a number of the stakeholders to continue manning the bases after the initial winter-over in 1940, Congress showed its feelings by drastically cutting funds and forcing the return of the expedition in 1941. By then the nation was being swept into World War II, and Antarctica was of little interest. No official reports of the expedition were ever compiled.

The war radically changed the world political scene. Europe had been devastated, the United States had emerged the primary military and economic power, and the Soviet Union was moving to consolidate its hold on the East. Antarctica became a component in a matrix of foreign policy issues in which the United States attempted to balance relations with Western allies and the growing threat of the Soviet Union.

In 1946–1947, however, before directions had become established, the United States staged the most massive human undertaking in the history of Antarctica—the U.S. Navy Antarctic Developments Project (code name Operation Highjump). Hastily established in seven weeks, the operation sailed on December 2, 1946, consisting of more than forty-seven hundred men, thirteen ships, and around two dozen airplanes and helicopters. The foray was primarily a military exercise of the U.S. Navy, flush with manpower following the war, and anticipating the possibility of conflict in the Arctic, where the Soviet Union was menacing Alaska, Canada, and Greenland. The logistical experience gained through this effort would give the United States distinct advantage in the events that transpired a decade later.

Task Force 86 (as it was called) operated a fleet that included two aircraft carriers, two ice breakers, a command ship with positions for fifty radio operators, and two task groups each consisting of a sea plane tender, a tanker, and a destroyer, the latter included for its speed in the event that a rescue situation arose. Each tender sent out two Martin Mariner seaplanes mounted with trimetrogon cameras to photograph the coastline of the continent. The cameras recorded three simultaneous images on nine-inch-square negatives, with views to the right, to the left, and vertically from the plane. By the time the two task groups completed their missions, they had recorded a vast area of the East Antarctic coastline, from 15° E to 170° E, much of it seen for the first time on these flights.

At the edge of the pack in the Ross Sea, the aircraft carrier *Philippine Sea* staged six ski-fitted R4D aircraft (military equivalent of the DC-3) that flew the seven hundred miles to Little America. Packed with radar and navigational instruments, in addition to trimetrogon cameras, they were also fitted for JATO (jet-assisted takeoff), a set of solid propellant rockets that fired at liftoff to boost the planes into the air from the short runway of the carrier and the sticky ice runway at Little America. The R4Ds flew numerous missions into unexplored areas of West Antarctica and the Transantarctic Mountains. A variety of tracked land vehicles also operated successfully out of Little America. Overall, Operation Highjump sighted an astonishing 1.5 million square miles of the continent and took more than seventy thousand aerial photographs.

The good-weather windows for flying from Little America were few and of short duration, so the crews and pilots were poised to jump when opportunity called. Such a break came on February 14, 1947. Shortly after midnight, with Admiral Byrd on board, the first R4D launched to the Transantarctic Mountains and flew straight for the Watson Escarpment, with the goal of delineating the Horlick Mountains. The plane flew up an outlet glacier to the east of the Scott (probably Leverett Glacier) and turned left (east) along the backside of the mountains. Falling oil pressure in one of the engines forced the plane to drop to lower elevations on the north side of the mountains, where it continued for more than seventy miles before turning back.

An hour later, a second R4D followed with a south-southeast heading, straight to the area of the Horlick Mountains viewed indistinctly during the flight from Little America in 1934. Turning east at the edge of the mountains, the plane followed the diminishing escarpment to a sort of termination, but out ahead another range rose with a steep shoulder emerging from between the ascending ice sheets, so they continued past it until they were forced to turn back due to hypoxia (lack of oxygen) of the crew.

With the first two planes safely returned to Little America, a second pair launched to the southeast, but stratus clouds limited the views to only a few dark summits protruding above the cloud banks. With the second pair of planes safely back, the freshly serviced first pair took off again, heading toward Mount Wade. Flying in tandem up the east and west sides of Shackleton Glacier (see Fig. 7.1), the planes turned right (southwest) at the head of the glacier, followed along the plateau side of the mountains, aiming for Mount Kirkpatrick, and then flew down Beardmore Glacier (see Fig. 4.11).

Clear weather held and another two planes launched for the South Pole, taking a route up Shackleton Glacier and then flying in along the 180° meridian. Byrd was on board, and at the pole dropped a cardboard box containing the flags of all of the members of the United Nations, then continued on for another ninety miles along the 0° meridian before turning back and crossing the Transantarctic Mountains between Beardmore and Shackleton Glaciers.

Aerial reconnaissance of Victoria Land began on February 17. The first three planes turned back due to either engine trouble or unfavorable weather, but the fourth completed a highly successful flight. The plane headed toward the mouth of Beardmore Glacier, veered west and flew over the Queen Alexandra Range, then turned north and followed along the backside of the mountains, hemmed in by clouds on the plateau side. As the plane rounded the west flank of Mount Markham, a huge convergence of cracked and fluid ice emerged, funneled off to the east into a steep defile, and streamed off into the Ross Ice Shelf (Fig. 7.2). Here were the headreaches of what Scott had named Shackleton Inlet, a spectacular outlet glacier that in its middle reaches dropped six thousand feet in a distance of less than twenty-five miles, in scale and intensity reminiscent of the middle reaches of Amundsen Glacier. In honor of Shackleton's ship, this major feature was named Nimrod Glacier. Continuing northward along the backside of the Churchill Mountains, the plane flew in over the Royal Society Range and then across the steaming summit of Mount Erebus before tracking back along the Barrier front to Little America (see Fig. 1.17).

With the next clear flying day, February 20, two planes flew west from Little America. The first flew in over Cape Murray to the north of Darwin Glacier and turned north, flying about one hundred miles along meridian 155° E to the head of Ferrar Glacier. From there it flew up the inland side of the mountains while the second plane flew in along the eastern side. Together these planes photographed the width of the Transantarctic Mountains as far north as Terra Nova Bay, for the first time observing and recording the system of ice-free valleys to the north of Taylor Valley.

The spaces of white dramatically disappeared from the map of the Transantarctic Mountains during Operation Highjump. Interior northern Victoria Land alone remained unsighted. However, no ground control existed for any of the newly photographed areas, so the production of accurate maps would not be possible.

Figure 7.2. The savage central portion of Nimrod Glacier constricts at Cambrian Bluff, the pyramidal peak on the left side of the image. Cape Wilson stands at the left rear. Scott's party made it to that spot on the last day of 1902 but was unable to reach ground due to the deep tears where Nimrod Glacier flows into the Ross Ice Shelf.

As a follow-up to Operation Highjump, the U.S. Navy staged Operation Windmill (1947–1948), so-called by the press because of its heavy reliance on helicopters. Task Force 39 consisted of two icebreakers, *Burton Island* and *Edisto,* whose primary mission was to secure ground control for the aerial photos of the previous year. The big ships pushed through heavy pack and, when open water permitted, nudged deep in toward shore, deploying the helicopters with crews who surveyed the terrain. The most successful activity was along the Wilkes Coast, resulting in the production of a set of middle scale maps. The icebreakers cruised down the Ross Sea to McMurdo Sound, lingered for a day visiting the historical huts on Ross Island, then sailed east along the margin of the ice shelf to Little America, without securing new ground control for any of the Transantarctic Mountains.

Setting the stage for postwar confrontation, Chile had matched Great Britain's claim to the Antarctic Peninsula in 1940, and then Argentina followed suit in 1943. Both subsequently sent forces to harass British bases on the peninsula. After the war, the United States attempted to bring stability to the region, as well as to exclude the Soviet Union. In 1948 it proposed an eight-way condominium for sovereignty in Antarctica, which included the United States and the seven claimant nations, Argentina, Australia, Chile, France, Great Britain, New Zealand, and Norway. The proposal was soundly rejected by all of the claimants. Then Chile introduced a countermeasure, the Escudero Declaration, proposing that all claims in the Antarctic should be frozen for five or ten years, for a cooling off period among the competing claimants, with all territories open to any expedition and with scientific research the only activity. Before negotiations could resume, however, the entire equation changed when the Soviet Union gave notice in 1950 that any further discussions of sovereignty must include the Soviets, owing to the "discovery" of Antarctica by Bellingshausen in 1820.

Over the next several years as the Cold War heated up, the United States grappled with Antarctic policy issues: containment, exclusion, and whether or not finally to make a formal claim. Meanwhile, a new force had been spawned in the postwar political environment. It grew quickly in strength and in reach and forever changed the politics of Antarctica. This political force was science.

Great strides had been made during the war in the fields of atmospheric physics, communications, and rocketry, with an international cast of elite physicists and engineers now poised to take their research to new levels. In 1950, at a dinner party at the home of James A. Van Allen, Lloyd V. Berkner proposed a Third International Polar Year (TIPY) to be held in 1957–1958, twenty-five years after the Second International Polar Year and coinciding with a maximum of activity in the sunspot cycle. The importance of the poles to the men in the room, who all studied some aspect of the upper atmosphere, was that the Earth's magnetic lines converged there. The measurement of solar activity in the ionosphere along magnetic lines that come to ground in the polar regions would provide data to test a model of Earth-sun interactions recently proposed by Sydney Chapman, the preeminent British geophysicist at the time, and the guest of honor at the Van Allen dinner party.

Berkner had been a radio technician on the first Byrd Antarctic Expedition (1929–1933), and after returning had joined the Department of Terrestrial Magnetism at the Carnegie Institute of Washington, where he directed studies in upper atmosphere geophysics. In 1932–1933 he had participated in the Second IPY. During the war he developed radar for the navy, and afterward he assumed advisory roles in Washington. As fast as Berkner could contact them, international scientific unions and societies endorsed his proposal for the TIPY: first the Mixed Commission on the Ionosphere, then the Union Radio Scientifique Internationale (URSI), the International Union of Geodesy and Geophysics (IUGG), the International Astronomical Union, and finally the parent institution of them all, the International Council of Scientific Unions (ICSU), which established a committee for the Polar Year in 1952. All these scientific groups were nongovernmental in their constitutions.

Additional scientific organizations, including the World Meteorological Organization and the International Association of Terrestrial Magnetism and Electricity, showed interest in participating and urged that the activity be extended to other regions beyond the poles. Chapman suggested that the TIPY be changed to the International Geophysical Year (IGY), and ICSU's special polar committee became the Comité Special de l'Année Géophysique Internationale (CSAGI). Members of this committee drafted letters requesting various nations to form national committees and to proceed with plans for research programs.

National committees and programs formed quickly. The U.S. program focused on upper-atmosphere physics and meteorology and planned one coastal and two interior stations in Antarctica, including the politically symbolic South Pole. The science budget became the responsibility of the fledgling National Science Foundation, founded by an act of Congress in 1950. As the IGY gained momentum, others joined the U.S. program, including marine geophysicists and glaciologists. Most of the logistical support for the scientists was to be provided by the U.S. Armed Forces, which also had their own strategic agendas for research and the siting of stations. The scientists seem to have found no contradiction in their using a military means to their civilian ends.

In the austral summer of 1955–1956, the U.S. Navy created Task Force 43 (nicknamed Operation Deep Freeze), under the command of Admiral George Dufek, whose orders were to sail to Antarctica and there build the new Little America V and the McMurdo Air Facility, and to begin preparations for the inland stations in Marie Byrd Land and at the South Pole. Britain, France, and the Soviet Union also built new stations that year. In 1956–1957, an additional fourteen stations were established by various nations, including Hallett Station operated jointly by the United States and New Zealand in northern Victoria Land. Located at the tip of Hallett Peninsula (see Fig. 1.5) in the midst of a large Adélie penguin rookery, this site was chosen after the nations abandoned plans to build the base at Ridley Beach at Cape Adare. The United States made the first aircraft landing at the South Pole on October 31, 1956, and construction of the Amundsen-Scott South Pole Station was under way. Meanwhile, an overland traverse of tracked vehicles provided the materials and manpower for Byrd Station in the middle of West Antarctica.

By the time the IGY officially opened on July 1, 1957, sixty-seven nations had fielded operations at more than one thousand stations around the globe—more than twenty-five thousand scientists and technicians and expenditures of more than $2 billion. Twelve nations operated fifty-five stations in Antarctica and on sub-Antarctic islands, with the "gentlemen's agreement" that they would share all data and not assert sovereignty. These included the seven claimant nations plus the United States, Soviet Union, Belgium, South Africa, and Japan. The U.S. program included studies of ionospheric physics, aurora, cosmic rays, geomagnetism, gravity, seismology, glaciology, meteorology, oceanography, terrestrial biology, and medical science. Missing from this list was geology.

In late 1956, before the IGY had even begun, CSAGI began a discussion about the possibility of extending the observations and occupying some of the stations for another year. The first day brought strong arguments on both sides of the question, but when on the second day of negotiations the Soviet delegation arrived and declared that all Soviet stations would remain operational, the recommendation quickly passed for IGY operations to continue for one more year under the rubric of the International Geophysical Cooperation program. The National Science Foundation established the U.S. Antarctic Research Program (USARP) to coordinate U.S. activities on the continent in January 1958, but it was not until 1960 that Executive Order, Circular A-51 was officially approved, setting forth the policies governing U.S. activities in Antarctica.

During 1959 the United States kept four stations open: South Pole, Hallett, Byrd, and McMurdo, with McMurdo becoming the base of operations and Little America shutting down because of the lack of a safe docking platform at the edge of the ice shelf upon which it was built. With the Soviet Union's intention of staying on the ice, however, pressure continued for the United States to maintain its stations after 1959, in particular the one at the South Pole. Sovereignty issues had been tabled during the IGY, but now the IGY had passed. Indeed, India had been urging the United Nations to consider Antarctica for peaceful purposes and the shared interests of all nations.

On May 2, 1958, in light of the evolving politics and following the course he had endorsed for the IGY, President Eisenhower invited the eleven other nations that had stations in Antarctica during the IGY to come to Washington to participate in a conference for the perpetuation of the ideals of cooperation established during the IGY, specifically along the lines of the Escudero Declaration. By June, representatives of the twelve nations were meeting on a biweekly basis in a room of the National Academy of Sciences in Washington. By early 1959 the main elements of the treaty were in place, and on December 1, 1959, it was signed at the Washington Conference. Immediately hailed as a milestone toward world peace, the Antarctic Treaty was then sent back to the respective governments for ratification.

On June 23, 1961, the last of the signatories endorsed the Antarctic Treaty. All territorial claims were put in abeyance for thirty years; the continent was to be used for peaceful purposes only; scientific research was to be the primary activity; there was to be no military presence except in support of scientific research; each signatory had to maintain a year-round base; all activity was to be open; information was to be shared. No mention was made in the treaty about mineral resources. Moreover, steps for other countries to accede to the treaty were established. The treaty was binding for thirty years and would continue in force after that period if none of the signatories withdrew.

By the thirtieth anniversary of the Antarctic Treaty in 1991, twenty-six nations held the status of "consultative party," and an additional fourteen had acceded to the treaty and its principles but were not conducting significant scientific research. At that time the consultative parties passed a far-reaching protocol toward protection of the Antarctic environment for at least the following fifty years, which included a moratorium on mineral exploration and exploitation, environmental impact statements for any activity in the Antarctic region, and guidelines for tourist operations on the continent.

In the fifty years since the IGY, Antarctic research has progressed with the issues of the day. Geologists have constrained reconstructions of Rodinia, the supercontinent predating Gondwanaland. Upper-atmosphere physicists and astronomers have built a telescopic array at the South Pole to measure the influx of neutrinos. Biologists monitor evolving ecosystems and marine food stocks. Glaciologists continue to assess the mass balance of Antarctic ice with the backdrop of disintegrating ice shelves, and they drill and sample layers to the bottom of the ice sheet with the intent of unraveling paleoclimate, all in the context of climate change and global warming.

During the IGY and in the years immediately following, numerous U.S. and New Zealand field parties fanned out from McMurdo Station and Scott Base, surveying the terrain and mapping the geology throughout the Transantarctic Mountains. The most high-profile operation in Antarctica during the IGY was the Commonwealth Trans-Antarctic Expedition under the leadership of Vivian Fuchs, which put in at the edge of the Filchner Ice Shelf. From there they staged a traverse with tracked vehicles across the ice cap to the South Pole, where they were greeted by a New Zealand party led by Edmund Hillary that had laid depots out from Scott Base in support of the British. The New Zealanders had found a suitable route for tracked vehicles through the Transantarctic Mountains via Skelton Glacier, in the reentrant to the southwest of Minna Bluff, which had gone largely unexplored by the early expeditions headed to the South Pole.

The New Zealanders spun off two dog sledge parties from the South Pole traverse to explore the plateau side of the Transantarctic Mountains as far south as Beardmore Glacier. The first worked along the uplands between Skelton Glacier and the western end of the Britannia Range, climbing prominent peaks and surveying the surrounding terrain before forging a passage down Darwin Glacier to the Ross Ice Shelf. The second party mushed southeast toward the head of Nimrod Glacier and then continued southeast along the flank of the Queen Elizabeth Range before returning across the plateau to rejoin the Trans-Antarctic Expedition.

New Zealand also fielded a party during the IGY that for the first time explored the interior of northern Victoria Land. The party of eight, consisting of four geologists, three surveyors, and a stores officer, all with considerable mountaineering experience, man-hauled up the Tucker Glacier from Hallett Station (see Fig. 1.5). These were the first persons to see beyond the imposing frontal wall of the Admiralty Mountains, first sighted by Ross's expedition more than one hundred years before. Tucker Glacier has carved a spectacular valley flanked by snow-clad buttresses of the Admiralty and Victory Mountains, whose summits soar to elevations above eleven thousand feet (Fig. 7.3).

The geologists and the surveyors mostly worked independently. The surveyors occupied a series of ten survey stations on peaks climbed on opposite sides of the glacier. The geologists did less climbing but traveled twenty miles farther up the Tucker, stopping at a distance sixty-six miles up from the mouth of the glacier. From their last camp they discovered two magnificent peaks, Mount Black Prince and Mount Royalist, in the heart of the Admiralty Mountains (see Fig. 1.6).

TRASH OR TREASURE?

As humans have explored the Antarctic wilderness further, their impact on it has become more problematic. For instance, in 1974 I traveled to Lake Vanda in Wright Valley across McMurdo Sound—a very beautiful place. My party had been stuck at McMurdo Station for five weeks waiting to be put into the deep field, and I had an acute case of cabin fever. So I ginned up a day trip by helicopter to Lake Vanda, which I had always wanted to see because of its unique, ice-free landscape. I was able to convince the National Science Foundation rep that we should examine the lake because one of the members of the party was a sedimentologist with an interest in the effects of algae in sedimentary environments, and we had heard that algae grew profusely in its frigid waters.

In late summer Lake Vanda is rimmed by a wide moat of meltwater, but when we flew there in November the moat was frozen solid (Fig. S.15). The ice around the lake was magnificently clear, becoming an ever-deepening field of blue and shot through with lacy white fractures (Fig. S.16). In the shallows, the clear ice revealed a blanket of algae wrinkled across the bottom.

As we hiked along the shore, marveling at the patterns in the clear ice and the great walls of Wright Valley that rose more than five thousand feet above us, we stumbled onto a collection of cans, rusting in a neat pile at the edge of the

Figure S.15. White ice of the permanently frozen portion of Lake Vanda is surrounded by clear blue ice formed following the summer influx of meltwater from glaciers at the head and mouth of Wright Valley. Note Vanda Station in the lower right of the photo.

Figure S.16. Fractures permeate the azure seasonal ice that rims Lake Vanda for ten months of the year.

lake. "How cool," I thought, "Who would have left these? Which year were they here?" I wondered whether I knew them personally or perhaps by reputation. I thought of taking one as a souvenir.

Then, about a hundred yards down the shore we came upon another pile. And then a couple hundred yards more, a third. By now I was disgusted. Images came to mind of the heaps of oxygen bottles at Everest base camp, or the calling cards of climbers under every rock on the path to the summit of the Matterhorn. Lake Vanda was trashed.

So where is the dividing line? To an archeologist, the dump at the mouth of a Paleolithic cave is a treasure trove of goodies, whether they be superbly crafted artifacts or the scraps of last week's big meal. The litter along Highway 61 has no less meaning about our current culture, but we find it rude and repulsive.

Since my visit to Lake Vanda in the 1970s, the U.S. and New Zealand programs have established a new policy: nothing is to be left in the Dry Valleys. Every ounce of waste must be bagged and flown back to McMurdo Station or Scott Base by helicopter. Furthermore, the two countries sent litter crews in to pick up the trash of the preceding decades. Today there are encampments in the Dry Valleys at various spots of scientific research, but when the research is completed, the researchers fold up their tents and depart without a trace.

For me, the dividing line between trash and treasure is the line between recent events and history. Tin cans in the Dry Valleys disgust me. But were I to find a sardine can dropped by Amundsen, a torn mitten worn by Scott, a broken ice axe left by Blackburn, or some of the debris scuttled by Byrd as he flew over The Hump on his way to the pole, such finds I would behold with awe and treasure as if a relic from the Holy Land.

Figure 7.3. Mount Trident, its 8,384-foot summit shrouded in stratus cloud, rises out of Tucker Glacier. The dark peak with vertical bedding that buttresses Mount Trident was named Trigon Bluff by the New Zealand survey party that climbed it during the International Geophysical Year, surveyed the lower reaches of Tucker Glacier from there, and left a surveying beacon.

The IGY also saw considerable activity by surveyors and geologists in the mountains and the ice-free valleys to the west of McMurdo Sound. Helicopters carried men into Taylor Valley and for the first time into Victoria Valley. In 1958–1959, Wright Valley was studied and mapped for the first time (Fig. 7.4).

A third spin-off of the Trans-Antarctic Expedition in 1957–1958 was a survey party consisting of two geologists, Bernard Gunn and Guyon Warren; a surveyor, F. R. Brooke; and a mountaineer–dog driver, M. H. Douglas. The party left Scott Base with two dog teams on October 4, 1957, and over the next 127 days traveled more than one thousand miles, literally circling the Dry Valleys. They worked up along the coast, closely following the route of David's party in 1907, with Gunn and Warren reaching as far north as Mawson Glacier, where they climbed Mount Gauss (see Fig. 3.8). Returning to Granite Harbour, the combined party traversed across the Wilson Piedmont and up Debenham Glacier (see Fig. 3.13). From a camp at Killer Ridge, Warren and Brooke took a three-day side trip to the south, where they climbed Mount Mahoney and surveyed from its summit. To the south the vista opened into a vast arena of bare rock (Fig. 7.5). This was Victoria Valley, photographed in passing from the air, but not until now witnessed and surveyed from the ground. Beneath the men a small frozen lake was fed from the disintegrating snout of Victoria Upper Glacier, a tributary in retreat pouring feebly from the plateau. From Killer Ridge the foursome sledged onto Mackay Glacier, passing the famil-

iar landmarks of Taylor's naming—Gondola Ridge, Sperm Bluff, Queer Mountain—as they sledged up to the plateau (see Figs. 1.14, 3.15).

After an airborne resupply at the head of Mackay Glacier, the party turned north and sledged all the way to Trinity Nunatak in the upper reaches of Mawson Glacier, thus completing two parallel traverses to 76° 30′ S, one along the coast and the other along the plateau (see Fig. 3.8). On December 15 the men were back at the depot at the head of Mackay Glacier, where they pushed south, glimpsing several times into the upper Victoria and Wright Valleys (Fig. 7.6). On December 26 they peered down the icefall where the men of Scott's party had slipped and fallen on December 4, 1903, on the return from their plateau traverse. From the head of Ferrar Glacier, Douglas and Warren flew back to Scott Base on January 16. Gunn and Brooke turned their attention to the backside of Mount Huggins, and relying on a bivouac partway up the mountain, put together a route to the summit of the second-highest peak in the Royal Society Range (see Fig. 2.5). Alas, the view on the far side was obscured by cloud. Following the descent they traversed down Skelton Glacier to the depot left by Hillary. There they were picked up and flown back to Scott Base on February 6, thus ending one of the most epic seasons in the annals of Antarctic mapping.

Figure 7.4. Upper Wright Valley, as viewed from Bull Pass, displays a classic U-shape, cut in the past by a glacier flowing into it from the East Antarctic Ice Sheet. This glacier has now retreated to the west of The Labyrinth at the very head of the valley (see Figs. 2.9, 7.6). The white feature in the bottom of the valley is frozen Lake Vanda, fed by meltwater of the Onyx River, flowing in on the left side of the image.

Figure 7.5. An icefront in retreat, Victoria Upper Glacier is the last vestige of a powerful glacier that once filled Victoria Valley and flowed through to the Ross Sea. This is the vista that Warren and Brooke looked into from the summit of Mount Mahoney, immediately out of the picture at the end of the prominent ridgeline that climbs to the left out of the picture. At the height of summer the melting glacier adds water to the frozen lake, which flows off into the heart of Victoria Valley.

Figure 7.6. The Labyrinth, an enigmatic landform at the head of Wright Valley, emerges from the edge of the retreating Wright Upper Glacier (see Fig. 2.9), shown at the bottom of the image. Lake Vanda appears beneath the set of hanging glaciers in the distance (compare Fig. 7.4).

DOUBLE PLUNGE

In the 1970s and early 1980s having a family member accompany a member of a research party to Antarctica was a fairly common practice. If the question of nepotism ever arose, NSF would counter that the lengthy time commitments to work on the Ice had led to many "Antarctic divorces." Giving a family member an opportunity to be a part of the program could thus be a preventive measure. Policies have changed, and now a family member who joins an expedition must have specific qualifications that justify the inclusion. In 1982–1983 I had the good fortune of being able to take my wife, Harriet, along as a field assistant, the best field assistant I've ever had. Before we left, Harriet's mother had made me promise that I would take good care of her daughter and bring her back safe and sound.

With a mission of surveying the basement rocks of the Dry Valleys, Harriet and I used helicopter close-support and small camps to visit and sample most of the region. One of our helicopter stops was on Mackay Glacier next to Queer Mountain. Rather than the typical ablation-pitted ice or sastrugi-adorned snow of a glacier, this area was smooth, hard ice of the kind found on a meltwater pond or lake that has frozen. The helicopter had set down about a quarter-mile from the rock, on a patch of rougher ice. Given that there were several inches of water at the edge of the rock, Harriet asked whether the ice was safe. I scoffed, "Sure, I've been on these things a hundred times before, and they are always frozen solid."

After we had collected samples and taken some pictures (Fig. S.17), we started walking back to the helicopter, single file, Harriet about twenty feet behind me. About one hundred yards out from shore, I heard a muffled swoosh, and a soft, but urgent, "Ed." When I looked around, there was Harriet, up to her armpits in water, leaning out of a large hole in the ice. At first glance it appeared that she was standing on the bottom, but in fact she was kicking frantically to stay afloat, with the bottom nowhere in sight. I lay down prone and reached out to her with my ice axe. She grabbed it and was able carefully to put one leg and then the other out onto solid ice. From there we slowly shuffled back to the helicopter side by side. Harriet was wet to the skin from her waist down, so the helicopter made a quick run directly back to McMurdo.

In the rush of the moment, this rescue was calm and successful. There was no doubt that we would get Harriet out of the drink and that everything would be okay. But in the relaxed security of the warm and pulsing helicopter, my adrenaline flooded in as I thought about what might have happened had the ice been thinner, had we both gone through, had the helicopter not been at hand and had we been some miles from camp. Only then did I get the shakes. "Harriet, what would I have told your mother?"

Figure S.17. Harriet stands on thin ice at the foot of Queer Mountain, minutes before her accidental plunge.

Later in the season we were dropped at Vanda Station, while the helicopter flew off to move another field party. Vanda Station had been built by New Zealand in 1967–1968 and had been operated continuously during the summer seasons, along with several years in the early 1970s when it had been staffed through the winter. Those who ran the station had started the Vanda Swim Club, which had gained considerable notoriety at the bases on Ross Island. Three rules governed induction into the club: a completely naked plunge into the frigid water that circled Lake Vanda in late December and January; total immersion; and witness by one of the "Vandals," or station personnel. That season the rules had been relaxed to allow some sort of footwear, since too many initiates had come out of the numbing waters with bleeding feet.

Harriet was keen to take a second plunge, so she stripped to her bunny boots and made her way to the shore wrapped in her parka. She was only the second female that season to join the club, and every Vandal on site turned out to witness the induction. Without hesitation, Harriet burst from her parka and charged into the water. When she was about thigh-deep, she threw herself under, with a rebound that gave the impression of a salmon leaping out a rushing waterfall (Fig. S.18). She intoned a cry somewhere between a scream and a gasp, and charged back up the bank to where I was holding her parka. After we retreated to the warmth of the station for tea and scones, everyone agreed that she had met the criteria splendidly, and I watched proudly as Harriet inscribed her name in the register.

Figure S.18. Harriet joins the Vanda Swim Club, January 1983.

The postscript to this story is that both the Vanda Swim Club and Vanda Station (see Fig. S.15, page 214) slipped into history as the victims of global warming. Every summer from the early 1970s, the Onyx River had delivered more meltwater to the lake than had ablated during the winter season. By the early 1990s it was clear that the station had only a few more seasons before it would be flooded, and so New Zealand began the sad task of dismantling the buildings and removing about ten tons of contaminated soil. By the end of the 1994–1995 season, not a trace remained of the base, which had for a quarter-century so admirably served the scientists who studied this polar oasis.

Three parties subsequent to the IGY conclude this account of exploration of the Transantarctic Mountains. In the second season of the IGY (1958–1959), a tractor train of three tracked Sno-Cats, each pulling a sled, lumbered out of Byrd Station to conduct geophysical and glaciological studies on the West Antarctic Ice Sheet. A similar party had worked the previous season in a big loop out to the Ellsworth Mountains and back, setting off explosives to sound the bottom of the ice sheet, digging pits and taking cores of the snow and ice, and surveying the ice sheet and the mountains. The routine was the

Figure 7.7. Shaded-relief map of the eastern Horlick Mountains. The thin broken line plots the route of the traverse party from Byrd Station as it approached the mountains. Bill Long and Fred Darling hiked sixteen miles to bedrock in the Wisconsin Range, as indicated by the yellow route. From a second camp at the northern margin of the Ohio Range, Long collected Devonian marine fossils, a first for the Transantarctic Mountains, from Mount Glossopteris. The ascent route is also shown in yellow.

same the second season, with refueling along the route by R4D aircraft, but this time the train headed south toward the Horlick Mountains, that last bit of outcrop where the East and West Antarctic Ice Sheets merge and the Transantarctic Mountains terminate. The party leader was Charles Bentley, a geophysicist and glaciologist probing the depths of the ice sheet. His second in command was Leonard LeSchank; together they manned the second Sno-Cat. The lead Cat was driven by Bill Long, a geologist by training but hired as a glaciologist, and his assistant, Fred Darling. The glaciologist was given the lead position in order to spot crevasses. To the end, the front Cat sported four twenty-foot whiskers composed of redwood four-by-fours with metal dishes wired ostensibly to detect hidden crevasses. In practice, it worked "about half the time." The rear Cat was shared by William Chapman, a topographical engineer from the U.S. Geological Survey, who surveyed ground control for future maps tied to aerial photos, and Jack Long, Bill's brother, who served as mechanic and kept the whole caravan running.

Bill Long had been on a reconnaissance flight earlier in the season and had spotted a major exposure of flat-lying sedimentary rocks in the last small range of the mountains. A goal of the traverse would be to touch ground at that spot for collecting. At a camp 339 miles out from Little America, the party was joined by photographer Emil Schulthess, delivered by R4D. At that point the train was stuck in a gnarly crevasse field on an ice stream that spilled across the single escarpment of what remained of the Transantarctic Mountains. The camp was 16 miles from the nearest rock, 16 miles of hard sastrugi and

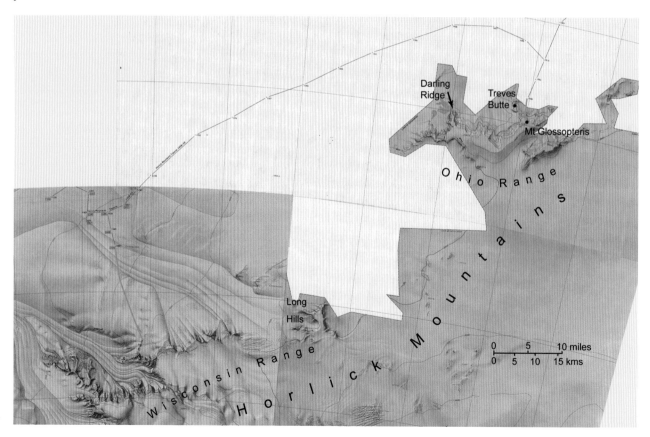

probably some crevasses, but Long and Darling were stout lads and the sun didn't set. Who needed to sleep anyway? Carrying no survival gear, they set off bound to touch rock. About eight hours later they reached the northernmost toe of outcrop at the mountain's edge (Fig. 7.7). The rock was granite, perhaps as expected. Long could see that farther on the granite rose in a fourteen hundred–foot cliff and was topped by sedimentary layers. These were the rocks that he most wanted to examine and collect, but climbing to them was out of the question. They had already pushed themselves to the limit. So they collected a few small specimens of the crystalline rock and trudged the 16 miles back to camp, arriving about sixteen hours after they had left.

The Sno-Cats managed to back out of the ice stream and pulled on east toward the small range Long had spotted earlier in the season. Each time the Sno-Cats stopped, Chapman would set up his theodolite and shoot a set of angles on bedrock points. In a week the party camped near the foot of a mesa whose sides were sheer cliffs of granite capped by a relatively thin section of sedimentary rock. The contact between the sedimentary layers and the granite five hundred feet above camp was the same unconformity exposed at the previous stop and throughout the Transantarctic Mountains (Figs. 7.8, 7.9).

Beyond the mesa, a more massive mountain rose several thousand feet above the surrounding ice. It was largely plastered over with snow, but a broad, smooth ridgeline with

Figure 7.8. Although the Ohio Range is nearly inundated by the East and West Antarctic Ice Sheets, it nevertheless displays the unconformity found throughout the Transantarctic Mountains, with Beacon sedimentary rocks overlying an erosion surface on older granite, as seen in this image of the escarpment to the east of Darling Ridge. Photo by Karl Kellogg.

Figure 7.9. With five hundred–foot cliffs of granite capped by sandstone, Treves Butte stands to the north of the main Ohio Range. Photo by Margaret Bradshaw.

a moderate slope climbed its face to the summit. The granite cliff was also there at the bottom, though not as high, and the rest of the mountain was built of a section of sedimentary rocks several thousand feet thick. The geologists were itching to see what the layers held, so Long, Darling, Bentley, and Jack Long hiked over to the outcrops under a sunny sky.

The party followed a drift around the granite cliff and came out onto a shelf of the sedimentary strata. Almost immediately the men found fossils—lots of them! Little marine shellfish, brachiopod bivalves, none any bigger than a quarter, but handsome, deeply grooved teardrops, and well preserved. Discoveries are serendipitous. The party had just discovered fossils like none other in the Transantarctic Mountains. Elsewhere, where the basal beds above the unconformity were known, they had been laid down by rivers meandering across floodplains on land. Here was clear evidence of deposition at the bottom of a shallow sea. Marine rocks like these, with fossils like these, had no counterparts.

Above the fossiliferous beds was a section of strange broken rocks, strewn through horizons of sand. Long had never seen anything like them, but he recorded their presence and moved higher. The section changed to mainly sandstones, which toward the top contained coal beds and, in some shalelike layers, unmistakable imprints of *Glossopteris,* the Permian plant fossil found previously in the Transantarctic Mountains by Scott and by Blackburn. But trumping everything were the petrified remains of tree trunks as much as two feet in diameter and ten feet long, littering the landscape at one horizon, looking as if some Permian catastrophe had leveled a forest.

With such a rich assortment of fossils and rock types, Long had to be selective in what samples the party returned to camp. From the summit, Bentley and Jack went directly back to camp, each carrying his share of rocks, while Long and Darling worked their way slowly down the ridge, collecting as they went. The weather had been clear during the ascent, but as the geologists descended, clouds drifted around the mountains, and they struggled before finding their way back to camp. Long had chosen his samples well, for the load of rocks the men hauled off the mountain that day would serve as the foundation of his master's thesis at Ohio State University. When he returned to the United States, he learned that the brachiopods were of a type known as *Terabratuloids* and Devonian in age; also, the strange broken horizons were tillites, deposited by glaciers, not of the present stage, but rather during an episode bracketed by the Devonian and the Permian. Although no analogous stratigraphic section was known throughout the Transantarctic Mountains, a close similarity existed to a section on South Africa. The ideas of continental drift proposed by Alfred Wegener early in the twentieth century had been stifled by the geological establishment. But geologists in the Southern Hemisphere had never let go of the idea of Gondwanaland, the southern supercontinent, because of how well it explained the commonality of stratigraphy and fossils found throughout the hemisphere's present-day continents and peninsular India. That Antarctica had a stratigraphy similar to southern Africa strengthened the notion of the connectivity of the components of Gondwanaland. Long's master's thesis became the basis for a proposal to collect, measure, and map throughout the eastern Horlick Mountains, subsequently funded by NSF for his Ph.D. dissertation.

By the IGY, Task Force 43 was Operation Deep Freeze III. The navy called on a variety of fixed-wing aircraft to support the program, including four of the old reliable R4Ds, ten ski-fitted DeHavilland UC-1 Otters for light support, and eight C-124 Cargomaster transport planes supplied by the air force. The Cargomasters were wheeled aircraft that landed on the smooth seasonal ice of McMurdo Sound and flew sorties out to the South Pole and Byrd Station, where they air-dropped supplies and fuel.

As activity in the Antarctic moved beyond the IGY, and Operation Deep Freeze III evolved into and then past IV, the Roman numerals changed to Arabic and years, so that the 1959–1960 season became Operation Deep Freeze 60. That season seven ski-equipped and one wheeled C-130 Hercules aircraft from the air force saw service for the first time. This plane, with its gaping rear cargo door and four powerful turbo-prop engines, could land at the South Pole or in the open field with a heavy load of cargo, and it was to revolutionize logistics to deep field locations for the ensuing years. By the following season (Deep Freeze 61) the navy had four Hercs of its own.

During this early period, aircraft fitted for trimetrogon photography made numerous flights along and across the Transantarctic Mountains, recording images for a comprehensive set of topographic maps, to be produced with two hundred–meter contours at a scale of 1:250,000. The Topographic Division of the U.S. Geological Survey had the job of producing the maps. What was needed before the maps could be drawn accurately was ground control. Survey parties had been in the field since the IGY, accompanying overland traverses—for example, Bill Chapman's work on the geophysical traverse to the Horlick Mountains in 1958–1959. They had also been placed close in to survey points by navy helicopters, but these were of limited range and especially elevation. The helos worked off the decks of icebreakers, and out of McMurdo, supporting field parties in the lower reaches of the mountains, but they didn't have the power to take men to high elevations where the full sweep of the mountains could be surveyed.

Chapman had heard about a new turboprop helicopter being developed by Bell that supposedly could land at thirteen thousand feet. He proposed to NSF that these be used during the 1961–1962 season to support a wide-ranging topographic survey based out of McMurdo Station. The foundation agreed, and requested support from the U.S. Army for the helicopter logistics. Two HU-1B helicopters (the first of the famous Huey line) were sent directly from the assembly line at Bell to Christchurch, New Zealand, where they were assembled and test flown, then taken apart, loaded onto C-124 Cargomasters, and flown to McMurdo Station.

The plan was bold: a series of remote camps supplied by fixed-wing aircraft from which the helicopters would operate, and two traverses—one south from McMurdo as far as Beardmore Glacier (Topo South) and one north from McMurdo as far as Cape Adare (Topo North), establishing ground control over more than eleven hundred miles of the Transantarctic Mountains. In a single season Chapman's party could conceivably master the terrain that had challenged an entire generation of heroic-era explorers.

In the first days of the season, the pilots made a series of test flights, shutting down at increasingly higher elevations on the sides of Mount Erebus and Mount Discovery. At the summit of Mount Discovery, at an elevation of eighty-eight hundred feet, one of the Huey's engines overheated and cracked when the pilot attempted to restart it. After installing a new engine (the only one they had in McMurdo), the pilots and crew succeeded in getting the helicopter off the mountain, but this failure resulted in new rules prohibiting the helos from shutting down at elevations above six thousand feet.

The procedure for surveying is to measure the location of a series of points out from an accurately positioned station. Three readings are necessary: a direction, a vertical angle or altitude, and a distance. The direction and vertical angle are measured by a theodolite, mounted with a telescope to see the "signal" the surveyor is targeting. Distance can be determined by triangulation—that is, by sighting on the same point from two different surveyed points. Chapman's party, however, used a Tellurometer, an Antarctic first for that instrument, which sent an electromagnetic signal that was bounced back by a reflector at the survey point, with the phase of the signal calibrated to distance. The precise positioning of stations at the start and the end of the traverse was accomplished by finding stars in the daylight sky with a telescope, rather than the more conventional celestial sightings.

The three engineers worked as two teams, placed by the two helicopters on peaks typically fifteen to thirty miles apart. Operating solo, the forward engineer would establish the site for a new station, set a tablet or benchmark in the bedrock (or use a pole if the site was snow covered), erect a signal composed of two three-foot-square orange cloth panels on a pole guyed to the ground, and set up and operate the remote Tellurometer unit. The two rear engineers would measure all the directions and vertical angles with the theodolite, with one man doing the sightings and the other recording the data. The rear team would also set up and operate the master Tellurometer unit, and would photoidentify the occupied station and all intersected peaks. If time permitted, they would also build a cairn. If a site was above six thousand feet, the helicopter would leave the engineers and fly to a lower elevation, where it would shut down and wait for a flash from a signal mirror or, if out of sight, would return at an appointed time. Then the forward engineer would fly to the next site while the rear engineers would occupy the previous forward site, repeating the entire routine and advancing the traverse.

The operation began on November 6, 1961, but an icy blast of wind brought clouds, obscuring the line of sight and numbing the men's fingers through their mittens as they attempted to make fine adjustments to the instrument with screwdrivers. The next day they added two stations—one at the summit of Mount Discovery, the other atop Mount Morning. From Mount Discovery the view swept from Ross Island, past White and Black Islands and Brown Peninsula, and across McMurdo Sound to the Koettlitz Glacier, decaying beside the foothills that climb to the massive wall of the Royal Society Range (see Fig. 1.17). In the opposite direction the view rounded the hooked end of Minna Bluff (see Fig. 4.1), then cut back along the shoreline of the snowy reentrant to the west, which had not been explored on the ground until the passing of the Trans-Antarctic Expedition in 1957–1958.

For the next several days, mixed weather mostly grounded the operation except for two more stations put in around Skelton Glacier, out at the limit of helo range from McMurdo. On November 11, with the weather clear and calm, the teams jumped into action early. By the time that they stopped working, it was 9:00 P.M., and they had occupied five new stations. From the last peak they flew down to the ice shelf to a prearranged location near Cape Murray, and there were delighted to find that the navy had already set up a camp, and even had rolled out the mattresses and sleeping bags in their Scott tents. An Otter and a navy HUS helicopter had delivered the gear, food, and fuel for the camp, along with the personnel to man it. Dinner was a "delicious meatbar stew" complemented by a two-ounce bottle of brandy for each man.

From this camp the Hueys would work south to the limit of their range, then the camp would be packed up and be moved by the navy helicopter to the next location. Throughout the traverse, fuel was delivered in fifty-five-gallon drums by a variety of aircraft, including Otters, R4Ds, and on two occasions by LC-130 Hercules. The Hueys refueled either at the base camp or at fuel caches set out by the navy helo. On November 12 the party had another excellent day, logging five more stations, the last of which was on the crest of the Britannia Range on a summit several miles to the west of the highest point, Mount McClintock (see Figs. 4.2, 7.13). The entire length of the mighty Byrd Glacier spread below the surveyors in a roughened ribbon of laminar flow, unwavering in the course it cut through the mountains. One false step from this station and a man could have plunged nine thousand feet down the south face of the Britannia Range, ice-bound and bristling with seracs. Barne and Scott had led parties to the mouth of Byrd Glacier nearly sixty years before, only to be barred from reaching rock by the savage crevassing where glacier meets ice shelf. That night the Topo South teams flew down to their next camp at Cape Selborne, close to the point where Scott had cached food in 1902 before turning his dogs south along the mountain front.

Over the next thirteen days the engineers occupied three more camps and surveyed eighteen additional stations. The stations followed close to the crest of the Churchill Mountains with views to peaks on both the plateau and ice shelf sides of the range (Fig. 7.10). Mount Durnford was particularly well placed, with its bold pyramidal mass set out in front of the main summit line of the Churchill Mountains, affording a view along the axis of the mountains (see Fig. 4.4). On Mount Albert Markham (see Fig. 4.5), thirty miles to the south, the second helicopter was attempting to land when it experienced a sudden loss of power and dropped about five feet before hitting the ground and sliding to a stop. Nothing was damaged, and the helo restarted without a problem, but the rear team was then shaken by a second incident. At the highest point on Mount Albert Markham, a large permanent drift hung at the top of a one thousand–foot cliff. As the engineers were setting up on this cornice, a loud crack resonated beneath their feet, engendering the apprehension that they were about to go sliding down the side of the mountain on this wedge of snow and ice. They scrambled under the hanging ice but could see no visible evidence of its letting loose, so they finished their job at the station above, even though every fifteen minutes or so another crack would detonate.

Another notable station was the one the surveyors established on Kon-Tiki Nunatak, the island in the middle of Nimrod Glacier (see Fig. 4.6). Surrounded by icefalls as chaotic as any in the Transantarctic Mountains, this fortress is assailable only by helicopter. To the south, Mount Markham reigned at the apex of a landscape in ascendance. Whereas the helicopters had landed on the highest summits in the Churchill Mountains, south

Figure 7.10. Satellite image of the central Transantarctic Mountains and a portion of the Queen Maud Mountains. The magenta dots are survey stations established by the Topo South party (Operation Deep Freeze 62), and the orange dots the stations of Topo East (Operation Deep Freeze 63). The red route plots Scott's 1902–1903 traverse, and the yellow route Shackleton's 1908–1909 traverse.

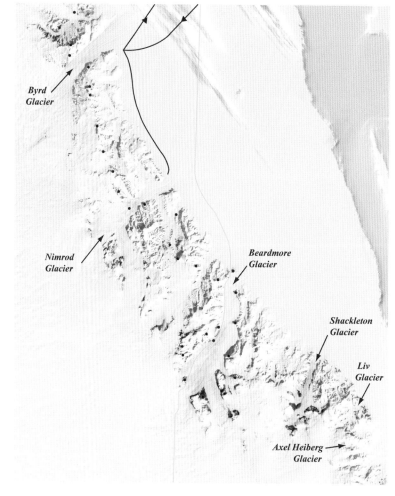

Byrd
Glacier

Nimrod
Glacier

Beardmore
Glacier

Shackleton
Glacier

Liv
Glacier

Axel Heiberg
Glacier

of Nimrod Glacier the Queen Elizabeth Range was simply too high for their operation. From Kon-Tiki Nunatak the line of stations skirted to the east of the Mount Markham massif and connected on south to a camp at an auxiliary weather station on the ice shelf about twenty miles out from the mouth of Beardmore Glacier.

At this point the survey party had to sit in camp for a week while fuel was being unloaded from the tanker in McMurdo. The delay did, however, give the men the opportunity to take astronomical observations for accurate positioning of the traverse. The LC-130 arrived on December 2 with sufficient fuel for the rest of the traverse. The following day they started up Beardmore Glacier, marking their first station on the summit of Mount Hope (see Figs. 4.9, 4.10), standing in the footsteps of Shackleton's party, where they looked up the great defile that would, if fortune smiled, bring them onto the plateau and thence to the South Pole. As the day progressed, the surveyors jumped to each of the prominent landmarks in the lower Beardmore drainage, first the whetted horn of Mount Kyffin, jutting into the glacier's flow, then Wedge Face, the next promontory to the south, then across the glacier to the top of The Cloudmaker (see Figs. 4.10, 4.11). At the end of the day, the party had established five new stations and was camped at a site set up by the navy crew at the southern end of The Cloudmaker, having traversed from Mount Hope to south of The Cloudmaker, covering in one day what it had taken Shackleton's and Scott's parties nine days each to traverse.

The following day was overcast so the party did not go out, but on December 6 the men completed a loop of seven stations around the upper end of the Beardmore drainage, twice crossing the tracks of Shackleton and Scott, landing at Plunket Point just downstream of Mount Darwin, then flying back to base camp on the ice shelf. After thirty-one days in the field, Topo South had completed its traverse.

At McMurdo, planning was soon under way for Topo North. Because the icebreakers were assigned to more pressing duties, the plan to use them in support of the traverse was scuttled in favor of perhaps two camps along the coast, supplied by Otters, with the Hueys sling-loading their own fuel drums out to remote caches. First, however, the party planned to work north from McMurdo to the limit of navy helo support. On December 17 the party launched, but the Tellurometer wouldn't work. The cause was a broken electrical contact, easily soldered back at McMurdo.

The following day weather was excellent, and the seasoned team pushed itself to its limits, surveying eight stations up to Mount Gunn, north of the head of Mackay Glacier (see Fig. 1.14). The traverse started at Brown Peninsula, the station where Topo South had also begun. The first stop was along the coast at Cape Bernacchi (see Fig. 3.8). From there the surveyors leaped to the summit of Mount Theseus, high on the crest of the Olympus Range separating Wright and Victoria Dry Valleys. These inner sanctums were unknown to the explorers of the heroic era, and had not been explored until the IGY. Each held a frozen lake in the depth, its heart fed by streams of meltwater emanating from glaciers at their opposite ends (see Figs. 7.4, 7.5, 7.6). From the Olympus Range, the surveyors hopped into the upper reaches of Mackay Glacier, the first outlet to the north of the Dry Valleys, and then into the southern Convoy Range for their last station of the day at Mount Gunn (see Fig. 1.14).

After enjoying a massive Christmas dinner McMurdo-style, the rested survey teams were ready for work on December 25. Flying directly to Mount Gunn, where they filled the tanks with fuel supplied by the navy helos, the teams started north along the lowest stretch of the Transantarctic Mountains, surveying up to a shoulder on the northern bank of David Glacier, where the broad outlet pinched and then released into the Drygalski Ice Tongue. This was David country, surveyed by Mawson in 1907 from points along the coast, en route to the Magnetic South Pole (see Figs. 3.8, 3.10). The mountains, as had been noted by Scott's first expedition, were blocky and tabular in shape. As the survey teams worked in the mountains, the navy flew Otters to Terra Nova Bay, where they landed on the sea ice at Cape Russell, built a camp, and depoted fuel. Separated by a ragged glacier front aptly called Hell's Gate, this camp was about three miles directly east of the beach on Inexpressible Island, where Priestley's party had wintered in 1911–1912. The surveyors might have pushed it past six stations that day, but katabatic winds poured down on them from the plateau, as is always the case in this area, freezing their hands and faces and keeping open the polynya at the southern end of Terra Nova Bay (see Fig. 3.12).

The next day clouds grounded the party, but the surveyors took the opportunity to measure their elevation relative to sea level. December 28 was bright and clear, but again the Tellurometer failed to operate, and again the reason was a broken contact. Since the camp was without the equipment to fix the circuit, an Otter had to be dispatched from McMurdo to take the instrument back, where it was promptly repaired.

Another Otter flew in the Tellurometer on December 30, and the party sprang into action. Its traverse took the men from David Glacier to the summit of Mount Melbourne (Fig. 7.11; see Fig. 1.12). The station at Mount Crummer looked down onto Backstairs Passage, where David's party had found a bypass onto the upper Larsen Glacier and gained the plateau (see Fig. 3.12).

At the start of New Year's Day 1962, the weather was clear. As the Hueys headed north to continue their traverse, the Otters launched from McMurdo to move the camp from Cape Russell to a site on Lady Newnes Bay. The first station was at the confluence of two glaciers, Aviator and Tinker (see Fig. 1.7), at a spot so narrow that the helicopter had to do a one-skid landing while holding power. The engineers hurriedly unloaded the helo under the full draft of its rotor, but when it had flown away they were drafted from another direction by severe katabatic winds, funneling down the glaciers and over their survey station. To make matters worse, in their hurry to unload the helo, they had left the windscreen on board. Unable to make their measurements under such gripping conditions, the men hastily built a wall of rocks about three feet high and knelt behind it while taking their readings with the theodolite.

The surveyors completed the next two stations with comparative ease as they worked the high country, but as the Hueys were beginning to run low on fuel, a radio message came that a fog bank was making it impossible to put in the camp on Lady Newnes Bay, and that the party should end its day at Hallett Station. The situation became critical when the helo crews failed to find a fuel cache that had been left for them by an R4D several days before. With almost no fuel left in their tanks, the helos landed to gain their bearings. At that moment the men heard an airplane flying overhead and radioed aloft their situation. As luck would have it, this was the same R4D crew that had left the cache,

Figure 7.11. (opposite) An icefall cascades down the steep northeastern wall of Priestley Glacier, flowing through its deep upper reaches toward the Ross Sea. Mount Melbourne towers above the landscape at the left rear of the image. From that vantage the Topo North engineers surveyed the region around Terra Nova Bay on December 30, 1961.

so the plane flew in under the cloud bank and proceeded to lead the helos to the fuel. More low clouds gathered as the teams completed their day's work, so again the helos descended into the fog, refueled, and then flew on to Hallett Station in whiteout conditions beneath the layer of cloud.

The following week both clouds and the men stewed around Hallett Station. During a few short breaks they were able to take six readings on stars to fix the northern end of the traverse. On January 8 the weather was marginal, and the helo party was becoming desperate. Throughout the day the helicopters dodged rolling clouds and completed four stations that brought the traverse to Hallett Station. The plan called for only five more stations from there, but Hallett Station remained socked in for the next two days. On January 11, three of the targeted stations came out of cloud. The engineers dispatched the first two with finesse, but as they were moving toward the third, the controls on the lead helicopter froze. While both pilot and mechanic pulled on the stick with all their might, the helicopter descended in a spiral and bounced to a safe landing on level snow. Inspection determined that meltwater had trickled in and then frozen around the steering mechanism, probably earlier that day at Hallett Station. The craft worked fine after the mechanic cleared the ice.

Perhaps this was an omen that it was time to cease this work. The season had certainly turned. Because of all the open water, the air was full of moisture, rolling up the glaciers and clouding over the peaks. So Chapman halted the traverse at Hallett Station, and the party flew back to McMurdo on January 12, after twenty-six days on the job.

The engineers of Topo South and Topo North completed a truly astonishing season, measuring a traverse distance of 1,510 miles by Tellurometer, and covering approximately 1,100 miles of the Transantarctic Mountains. They occupied seventy stations, set fifty stationary tablets, and intersected 113 points, overall controlling an area of 100,000 square miles for the 1:250,000 scale topographic maps that were already rolling out of the U.S. Geological Survey. Never before nor since has a party in a single season seen so much of the Transantarctic Mountains in such detail, encompassing such a spacious sweep of history.

Flush with their success, the U.S.G.S. and the army were back in Deep Freeze 63 with a team of surveyors and the two HU-1Bs. This season the leader of the party was Pete Bermel, a topographer who had been to the Ice two years earlier on a motor toboggan traverse to the Horlick Mountains and the solitary Thiel Mountains, poking from beneath the ice 125 miles to the east of the Ohio Range. The plan for 1962–1963 was to extend the traverses of the previous season, first by doing a loop to complete northern Victoria Land (Topo West) and then extending in the other direction from Beardmore Glacier to the end of the Ohio Range (Topo East).

The logistic plan that had served so well in the previous season was basically followed again, with the navy maintaining and moving the camp and resupplying with fixed-wing aircraft. The Tellurometer, however, was replaced with an improved model that could both send and receive a signal, so two pairs of topographers worked on alternating peaks, leapfrogging each other rather than having to occupy each station twice, as had been the case the previous season. Topo West started from Hallett Station and worked west along the northern, ice-choked coastline of northern Victoria Land, the one that Ross had surveyed in late February 1841 (see Figs. 1.5, 1.9). Beyond Cape North the traverse carried along the low coastal ranges as far as 158° E to a feature known as the Wilson Hills. From there the party backtracked to the lower reaches of Rennick Glacier and took the traverse south along the elongate bluffs and ranges of central northern Victoria Land, closing at the station on the shoulder of Mount Murchison that Topo North had established the season before.

Topo East would complete the Transantarctic Mountains survey. Starting at the Mount Kyffin station on the south side of Beardmore Glacier, the first part of the traverse crossed the drainage basin of the Ramsey Glacier, heading in the highlands between the outlets of Beardmore and Shackleton Glaciers (see Fig 7.10). On the south side of Shackleton Glacier, the party established a station at the summit of Mount Munsen (see Fig. 7.1), the peak immediately to the north of Mount Wade, from which the team surveyed the stretch of mountains between Shackleton country and Byrd country that had remained unknown until the Condor fly-by in February 1939, during the U.S. Antarctic Service Expedition. From there Topo East hopped across Liv Glacier and occupied a station at Mount Balchen, the high, eastern shoulder on the Fridtjof Nansen massif (see Figs. 5.3, 5.5). From this vantage the surveyors looked directly down onto the icefalls that Amundsen's party had mastered in November 1911.

The next station was the east corner of Breyer Mesa (see Figs. 5.10, 5.13, 6.2), looking down onto the savage central section of Amundsen Glacier and across to the grouping of high summits centering on Mount Astor. From there the parties hopped to Mount Bowser, at the southwestern end of the cirque, where the Souchez Glacier originates (see Fig. 6.2). At the head of the cirque about two hundred feet higher and only three miles around the summit ridge stood Mount Astor, the highest peak in the Hays Mountains, and around from that on the opposite side of the cirque, Mount Crockett, about five hundred feet lower. The station commanded a vista across the entire middle and upper reaches of Scott Glacier. This was Blackburn country, with its spectacle of granite and its southern exposure. The campsite for work in this area was located on Bartlett Glacier, about thirty miles to the south.

At the same time a geological party was camped near the head of Scott Glacier on the snowfield beneath the icefall joining Mount Weaver and Mount Wilber (see Fig. 6.9). It was there to extend the stratigraphy that Blackburn had measured at Mount Weaver into the surrounding ranges, and to see whether that could be correlated with the sedimentary sequence in the Ohio Range. Based in a small Jamesway (a portable, modular, Quonset-style hut with canvas top and wooden floor), the party of four from Ohio State included geologists George Doumani (leader) and Velon Minshew (Ph.D. student) and field assistants Courtney Skinner and Larry Lackey. Doumani had been hired during the IGY as a geophysicist working on Sno-Cat traverses out of Byrd Station, and had crossed paths with Bill Long in November of 1958 as an opportunistic passenger on the R4D that delivered Emil Schulthess to the party in the Horlick Mountains. He continued traversing in West Antarctica in 1959–1960. The next year, he signed on as the paleontologist with a project from Ohio State built around Long's discoveries in the Ohio Range two

234 Figure 7.12. The conduit along which lava erupted at Mount Early is displayed in this image from close to its summit. The yellow rock on the right is palagonite, a fragmented glass that was produced at an early stage of the eruption when the magma poured into the overlying ice sheet and shattered. As the dark magma on the left rose through this quenched glass, boiling of water at the interface between the yellow and the black produced fluidization channels that shot into the as-yet-unconsolidated palagonite.

years before. Their camp, put in by R4Ds, included a Jamesway structure and used snowmobiles for the first time to visit nearby outcrops of rock.

During the 1962–1963 season at the Mount Weaver camp, the party had had continual difficulties in keeping their snowmobiles running and was eagerly awaiting the promise of a helicopter to assist in the geological mapping. The one outcrop somewhat beyond Mount Weaver, which the party had reached by snowmobile, was Mount Early, the conical peak fifteen miles to the south that Blackburn's party had spotted on December 10, 1933 (see Fig. 6.10). Much to the geologists' surprise, Mount Early was a small volcano, the southernmost volcano on earth. As they climbed to its summit, fragmented magma in the outcrop gave the appearance that they were actually ascending up along the volcano's eroded throat (Fig. 7.12).

During the first week in January, three helicopters visited the geologists' camp. First Pete Bermel and Captain Frank Radspinner, the army commander, arrived to make contact and determine the geologists' goals; the second helo brought in a drum of fuel; and the third, piloted by CWO John D'Angelo, arrived ready to fly at the geologists' bidding. Their first request, obviously, had to be Mount Howe. From the summits of Mount Weaver and Mount Early, one could clearly see that the mountain was built of sedimentary layers (see Fig. 6.11). What better geological justification to be the first humans to set foot on the southernmost rock on the planet? The Huey landed at the foot of the cliff, where almost immediately the geologists found *Glossopteris* in the strata. When the mountaineers pulled out the ropes to scale the cliff, D'Angelo smiled at their machismo and delivered the men to the summit of Mount Howe a couple of minutes later. The delighted party stopped several more times at outcrops on the flight back to the Mount Weaver camp, accomplishing in four hours what they had planned to do in three weeks with their unreliable snowmobiles, lithely soaring over an almost continuous crevasse field between stops.

The helicopter stayed with the geologists for several more days, visiting in touch-and-go reconnaissance fashion most of the outcrops in the upper Scott Glacier area, before rejoining the Topo East traverse that had been busy with a series of stations that looped Scott Glacier, including Mount Saltonstall, to the north of Mount Weaver; Mount Grier in the La Gorce Mountains; and Mount Gardiner, abreast of Mount Blackburn at the confluence of Bartlett and Scott Glaciers (see Fig. 6.2). The next three stations were cast on the top of the Watson Escarpment, surveying the lowlands beginning at the backside of the Gothic Mountains, past the narrow breach where Leverett Glacier breaks through from the plateau, and around from there into the drainage of Reedy Glacier (see Fig. 5.20), the last major outlet crossing the Transantarctic Mountains before the East Antarctic Ice Sheet merges with its West Antarctic counterpart beyond an irregular escarpment that ends at the Ohio Range.

Having reached the distal termination of the mountains, Topo East backtracked to Mount Weaver, where the full detachment of ten men and two helos invaded the geologists' camp on January 24. They were there ostensibly to give further assistance to the geological party, but they were positioning themselves to leap from land's end and alight at the South Pole. The geologists were served admirably by the army pilots, while Radspinner argued with the navy about whether and how he would fly his Huey to the pole. Eventually the geologists collected all the rocks they could carry and were flown by R4D back to McMurdo, leaving the helo detachment in the field. The navy finally granted permission for the army to make the first helicopter flight to either of the poles, but only on the condition that one of their own LC-130 Hercules navigate to the South Pole, leading the army helicopter with the contrail of its engines. With permission granted, Radspinner piloted his HU-1B the 176 miles from Mount Howe to the South Pole on February 4, 1963, making that historic flight the capstone to two seasons of Antarctic exploration that have never been matched in scope.

Within two years every 1:250,000 quadrangle of the Transantarctic Mountains was in print. By this benchmark, the geographic exploration of the Transantarctic Mountains was complete. By 1965 every spur on every ridgeline, every little tributary valley, was mapped and published in beautiful detail. It was a wondrous accomplishment, one could say a work of art, a rendering in some ways more real than the subject it represented.

But at the same moment in history that we had reached the limit of global exploration, a vast new arena opened. It happened on October 4, 1957, during the first months of the International Geophysical Year when a shining object named *Sputnik* streaked across the heavens signaling to the world that the age of rocket-borne exploration had begun. In the ensuing half-century, an ever-increasing array of satellites has looked down on Earth, and probes have been launched to all quadrants of the solar system. Within a decade we had seen the dark side of the moon and shortly thereafter had landed lunar astro-

Figure 7.13. From a drift on the Lowry Massif, a lone figure surveys the scene. Across Byrd Glacier stands Mount McClintock, the highest point in the Britannia Range.

nauts who brought back samples of rock, rocks dated with ages older than any on Earth. In the ensuing decades we have imaged the moons of Jupiter, landed rovers on Mars, captured the dust of comets, and focused our orbiting telescopes on the very boundaries of the universe.

In concert with the revolution in rocketry has been the revolution in computing. From the giant computers with names like Eniac and Univac, we have created ever-smaller and more powerful machines, which, when linked to the Internet, reach into every aspect of our daily lives. Global positioning systems tuned to an array of orbiting satellites can tell us to within millimeters the place we stand, at a price that makes it an affordable option on any new SUV, and at an accuracy beyond the dreams of surveyors of the IGY. Today anyone with a personal computer can go to GoogleEarth and count the swimming pools down the block, or count them as easily in Katmandu or Timbuktu.

Terrestrial exploration of Earth has reached its logical limit. The frontiers of the unknown have shifted to the depths of the oceans and the surface of Mars, to the intricacies of genetic codes and the underpinnings of subatomic particles, to the very boundaries of the universe. Those who probe the mysteries of these realms are a privileged few, reporting back to nonscientists the wonders that they have beheld. In this way, perhaps, they are no different from the geographical explorers of yore (Fig. 7.13).

Epilogue

A spur of rock exists in the Duncan Mountains reaching down from the main ridgeline of that foothills range. It is an ordinary spur flanked by equally ordinary spurs on either side. The sort of nothing-special place you'd never choose to climb if you had journeyed so deeply into this deep-field space—unless perhaps you were a geologist.

We were out that day chasing a fault that ran along the backside of the summit ridge, coming down the crest of that ordinary spur—black, fine-grained schist dipping into the mountain, less frost-shattered perhaps than other spurs down the line—descending with a sure-footed lope and the click of our ice axes hitting hard rock. The day was windless. We worked in shirtsleeves, hats, and gloves.

Imperceptibly my cadence slipped into my consciousness. Each thump of my boot and tap of my axe was being mimicked. My immediate thought was that the boys up behind were messing with me, marking their footsteps to mine. But that certainly wasn't the case. As the other three ambled down to my spot, their footsteps and their voices came echoing softly from the spur to the east. We whooped and were answered back by a whoop, a single whoop, then silence. We made jungle noises and shouted names and thought that the whole thing was pretty cool. Then we shouldered our packs and headed farther down, hoo-hooing as we went.

Before we had gone twenty paces, the echo had become noticeably louder. Delighted, we plunged deeper into the auditory node. Soon the mountain was booming back at us, so loudly that we lowered our voices. We had slipped out of the fun house and into the nave. Our every movement was resonating. The air had become tangible. We tiptoed

Figure E.1. Latter-day explorers drive snow-mobiles and sleds over blue ice toward an unnamed peak on the west side of Scott Glacier.

down to a level where the returning sound was not diminished, where our softest whispers came back with clarity and nuance, almost as if they were being amplified.

Who could have imagined such a singularity on such an ordinary spur? With proper measurement, one could describe the curvature and focus of the terrain and explain the echo in elegant mathematical terms. But would that reduce the mystery of its being, the coincidence of forces that created the resonance, the unlikelihood of our entering that vibrant space? The mountains will denude, the configuration will be altered, the echo will be lost. But once at least a passing footstep chanced to penetrate the secret. How many other echoes wait on lonely ridgelines? How many other seekers will stumble on them there?

Nature's mysteries permeate the landscape, pattern it at every scale. It is a matter of positioning one's self, of being watchful . . .

And so we return to the mountains again,
back to this splendid disruption
at the end of the Earth,
with yearnings for simplicity,
for rigor,
for reunion of body, Nature, and mind,
for mastery of space and inklings of time,
the thin, clean air, the vistas,
and near the end, perhaps, the glimpse of an order,
an underpinning to the complexity,

the randomness of pattern,
the visual field elongate in perspective,

    the deep field
      wherein brilliant arrays of edge and flow
      pulse in the spectral atmosphere—

where memories of warmth and life are lost
or never dared to be,
where emptiness hangs on the plains of ice,
where process pervades,
great, sweeping process, transforming, enhancing,
utterly unto itself.

Into this emptiness the mind reaches,
fixes on a point of reference,
expands to encompass the horizon.
The body shivers.

Silence abides.

Figure E.2. Rock, ice, and cloud: the three elements of the Transantarctic Mountains create a stark image at the Mount Borcik massif, midway along Scott Glacier.

Appendix 1
The Rock Cycle

Despite the wondrous variety of rocks that geologists have identified in the crust of the Earth, we classify them into only three major types: *igneous, sedimentary,* and *metamorphic,* each of which forms through the action of geological processes on other rocks. All rocks (except for coal and some volcanic glasses) are composed of minerals, and all minerals are composed of atoms in specific proportions, arranged repetitively in specific crystal structures. The transformations of one rock type to another may be conceptualized using a diagram called the Rock Cycle (Fig. A.1).

The most fundamental rock type is igneous rock, which forms when minerals crystallize from melted rock, or *magma.* Magmas form deep in the Earth by the melting of other rocks, and being liquid and of a lower density than the rocks from which they melted, they rise buoyantly. If magma cools and crystallizes before it reaches the surface, the resulting igneous rock is said to be *intrusive* or *plutonic.* If the magma reaches the Earth's surface and erupts, the resulting rock is said to be *extrusive* or *volcanic.*

At the Earth's surface the processes of *weathering* and *erosion* continually wear away whatever rocks are exposed, be they igneous, metamorphic, or sedimentary. The rocks are broken down into individual mineral grains or rock fragments, or are dissolved into water, and are transported by currents of water or wind to places where they are deposited or precipitated as *sediments.* As sediments accumulate, those at the bottom of the sequence are subjected to increasingly higher temperatures and pressures, which cause compaction and cementation that transforms the sediment into *sedimentary rock.*

If sedimentary rocks are buried to even greater depths, the increased temperature and

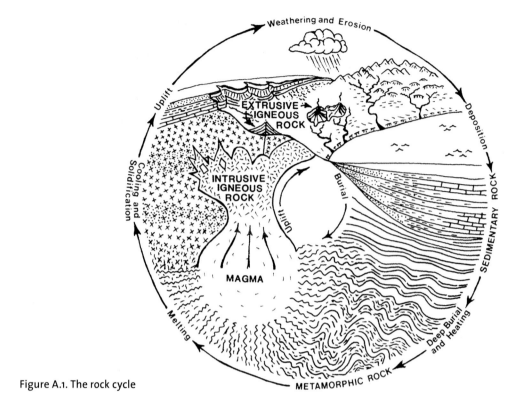

Figure A.1. The rock cycle

pressure will cause the breakdown of some minerals and the growth of others. This solid-state transformation or *recrystallization* of one suite of minerals to another, which takes place in the solid state, is called *metamorphism,* and the resulting rocks are *metamorphic.* If metamorphic rocks are subjected to high enough temperatures, they will melt, and when the resulting magma cools, igneous rocks will have formed, completing the Rock Cycle.

Appendix 2
Geologic Time

Geologists have two ways of telling the age of a rock, relative dating and absolute dating. Absolute dating tells you an age in years, or more typically millions of years. The dates are derived by measuring radioactive isotopes with mass spectrometers. The technology for isotopic dating is a development of the twentieth century, so none of the early Antarctic explorers knew more than an approximate age for any of their rocks. They relied instead on a variety of relative dating principles, which allowed them to determine the relative age of two rocks that were in contact with each other.

The first principles, called superposition and original horizontality, were put forth by Nicholas Steno (1638–1686), physician to the Medici of Renaissance Florence, who also studied the rocks of the Tuscan region. According to the principle of superposition, when one observes a sequence of sedimentary rocks, the oldest layer or bed is always on the bottom, and the layers above become systematically younger the higher you go. According to original horizontality, sedimentary layers are horizontal when they are deposited, and if one finds a sequence of sedimentary layers that are not horizontal, then they were tilted or deformed by a later geologic event.

A century later James Hutton (1726–1796), a gentlemen farmer and businessman, was doing his own inquiries into the nature of the rocks in his native Scotland, where he recognized two more ways to tell relative geologic time. The first is the principle of cross-cutting relations, according to which an igneous rock that intrudes another rock is younger than the rock that it intrudes. The same is true of a fault that breaks and displaces rock; it will be younger than the rock that it cuts across.

Hutton also recognized an important geological feature called an unconformity, which is essentially an ancient erosion surface preserved beneath a sequence of sedimen-

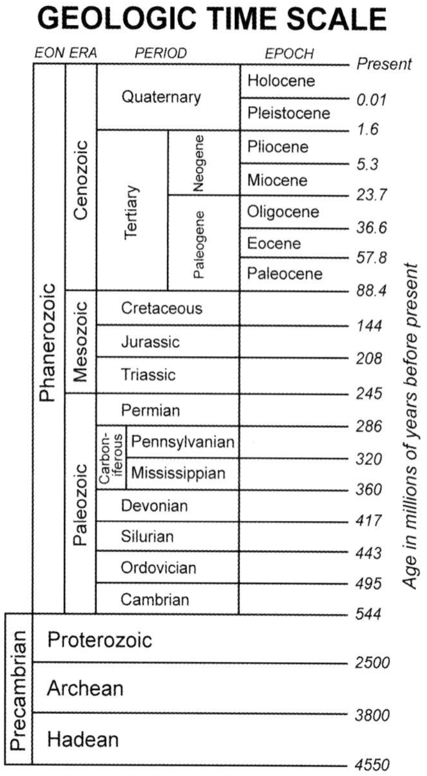

GEOLOGIC TIME SCALE

Figure A.2. Geologic timescale

tary rocks. The rocks below an unconformity may be of any type, igneous, metamorphic, or sedimentary, and if they are sedimentary, they typically have been deformed. In fact, rocks below an unconformity have typically been involved in an episode of mountain building, a complete revolution of the rock cycle, with rocks formed deep in the crust brought to the surface by uplift and erosion, followed by the start of a new sedimentary sequence upon that surface producing the unconformity.

William Smith (1769–1839), a British engineer of roads and canals, made perhaps the most profound discovery of all relating to telling geologic time. Having collected fossils for many years throughout Great Britain, he came to recognize that a great variety existed among fossils, and also that the order or succession of these varying fossils in sedimentary rocks was always the same, no matter where you found them. When Smith's principle of fossil succession is coupled with superposition, then the fossil record itself becomes a timepiece, with the oldest fossil occurring in the lowermost layer and younger fossils occurring successively higher up.

The fossils that Smith reported by no means covered all of geologic time. For the first half of the nineteenth century, the challenge for fossil hunters was to measure and record all of geologic time as represented by the fossil record. The result of this intense labor was the geologic timescale (Fig. A.2). Each of the periods of geologic time was based on a specific sequence of fossils, and each was given a name. Although it would take another century before absolute ages were added to the boundaries of the Geologic Timescale, the fossil record provided an exquisitely detailed means of calibrating the age of a rock if that rock indeed contained a fossil. This is why all of the early Antarctic geologists were so keen to find fossils.

Glossary

Ablation Removal of ice and snow by evaporation and/or melting.

Ablation cups Bumpy texture on ice surfaces produced by differential evaporation as the sun shines from different angles in the sky.

Basalt A black–to–dark gray, fine-grained igneous rock of extrusive or volcanic origin. A common rock type in island volcanoes, it also composes all oceanic crust.

Batholith A body of intrusive igneous rock of regional scale, composed of many individual plutons.

Contour interval The difference in elevation between two contour lines.

Contour line A line on a topographic map connecting points of equal elevation.

Dike A tabular pluton that intrudes across layers of rock.

Dolerite A black–to–dark gray, fine-grained igneous rock of shallow, intrusive origin. Has the same chemical and mineralogical composition as basalt. *Dolerite* is a British term, first applied in the Transantarctic Mountains by Ferrar. The equivalent term in the United States is *diabase*.

Finnesko Boots made of reindeer hide with the fur on the outside.

Glacier A stream of flowing ice.

Glissade A sliding descent of a snow slope in a standing or crouched position, using the tip of an ice axe for balance and braking.

Gneiss A metamorphic rock of high temperature and pressure in which dark and light-colored minerals separate into bands.

Granite A light-colored, coarse-grained igneous rock of intrusive origin. Granite, the most common rock type in continental crust, forms typically in the deep regions of active mountain belts.

Headreach The uppermost portion or place of origin of a glacier or a stream.

Hoosh A soup made of pemmican, sometimes dried biscuit, and water.

Icefall A steep flow of ice producing broken blocks.

Katabatic wind Dense, gravity-propelled wind, produced in Antarctica when the ice sheet is colder than the air above it. The air cools, contracts, and becomes denser. Drawn by gravity, this heavy air pours downslope across the ice sheet, and in places funnels into the headreaches of outlet glaciers.

Land blink A bright atmospheric phenomenon at the interface of sea and sky produced by radiance from snow-clad mountains beneath the horizon.

Moraine A ridge of till. A moraine may be lateral, at the side of a glacier; medial, as a band within a glacier downstream from a point where two glaciers merge; and terminal, at the end of a glacier.

Névé A snowfield of low relief.

Nunatak An island of rock surround by ice.

Outlet glacier A glacier that flows from the East Antarctic Ice Sheet across the Transantarctic Mountains into the Ross Sea or Ross Ice Shelf.

Pack ice Broken blocks of sea ice, tightly or loosely packed.

Pegmatite An intrusive, igneous rock, generally of small size, with extremely coarse-grained minerals.

Pemmican Ground, compressed, dried meat and fat, possibly mixed with vegetables and/or berries. The standard trail food of the explorers of the heroic era.

Piedmont glacier A glacier that flows from a valley or highland out onto a relatively low-relief terrain. Piedmont glaciers typically expand to a broad, lobate form on the piedmont.

Pluton A body of intrusive igneous rock.

Sandstone A sedimentary rock composed of sand-sized (and often smaller) particles.

Sastrugi Wind-blown patterns in snow.

Schist A metamorphic rock of intermediate temperature and pressure with shiny, reflective layers due to the growth of mica flakes.

Scree Rock debris on the side or base of a slope.

Serac A block of ice in an icefall.

Shale A sedimentary rock composed of mud or clay.

Sill A tabular pluton that intrudes parallel to layers of sedimentary rock.

Talus Rock debris on the side or base of a slope.

Theodolite A surveying instrument used for measuring horizontal and vertical angles.

Till Unsorted sedimentary debris (clay-size to boulder-size) transported and deposited by a glacier.

Unconformity An erosion surface in the rock record overlain by layers of sedimentary rock.

Bibliography

Amundsen, Roald. *The South Pole*. New York: Lee Keedick, 1913.

Armitage, Albert B. *Two Years in the Antarctic: Being a Narrative of the British National Expedition to Antarctica*. London: Edward Arnold, 1905.

Bernacchi, Louis C. *To the South Polar Regions: Expedition of 1898–1900*. London: Hurst and Blackett, 1901.

Bertrand, Kenneth J. *Americans in Antarctica, 1775–1948*. New York: American Geographical Society, 1971.

Borchgrevink, Carsten E. *First on the Antarctic Continent: Being an Account of the British Antarctic Expedition 1898–1900*. London: George Newnes, 1901.

Bull, Henrik J. *The Cruise of the "Antarctic" to the South Polar Regions*. London: Edward Arnold, 1896.

Byrd, Richard Evelyn. *Little America: Aerial Exploration in the Antarctic; The Flight to the South Pole*. New York: Putnam's, 1930.

——. *Discovery: The Story of the Second Byrd Antarctic Expedition*. New York: Putnam's, 1935.

Crary, Albert P. "International Geophysical Year: Its Evolution and U. S. Participation." *Antarctic Journal of the United States* 17, no. 4 (1982): 1–6.

David, T. W. Edgeworth, and Raymond E. Priestley. *Glacialogy, Physiography, Stratigraphy, and Tectonic Geology of South Victoria Land, with Short Notes on Paleontology by T. Griffith Taylor*. London: W. Heinemann, 1914.

Doumani, George A. *The Frigid Mistress: Life and Exploration in Antarctica*. Baltimore: Noble House, 1999.

Drewry, David J., and Roland Huntford. "Amundsen's Route to the South Pole." *Polar Record* 19 (1979): 329–336.

Ferrar, Hartley T. *Report on the Field Geology of the Region Explored during the "Discovery" Antarctic Expedition, 1901–4*. London: Museum of Natural History, 1907.

Gould, Laurence M. *Cold*. New York: Brewer, Warren and Putnam, 1931.

Herbert, Walter W. "The Axel Heiberg Glacier." *New Zealand Journal of Geology and Geophysics* 5 (1962): 681–706.

McCormick, Robert. *Voyages of Discovery in the Arctic and Antarctic Seas, and Round the World: Being Personal Narratives of Attempts to Reach the North and South Poles.* 2 vols. London: Sampson Low, Marston, Searle, and Rivington, 1884.

Paine, Stuart D. *Footsteps on the Ice: The Antarctic Diaries of Stuart D. Paine, Second Byrd Expedition.* Edited with an Introduction by M. L. Paine. Columbia: University of Missouri Press, 2007.

Priestley, Raymond E. *Antarctic Adventure: Scott's Northern Party.* London: T. Fisher Unwin, 1914.

Prior, George T. *Report on the Rock Specimens Collected during the "Discovery" Antarctic Expedition, 1901–4.* London: Museum of Natural History, 1907.

Quartermain, Leslie B. *South to the Pole.* London: Oxford University Press, 1967.

———. *New Zealand and the Antarctic.* Wellington: A. R. Shearer, Government Printer, 1971.

Ross, James C. *A Voyage of Discovery and Research in the Southern and Antarctic Regions during the Years 1839–43.* 2 vols. London: John Murray, 1847.

Schulthess, Emile. *Antarctica.* New York: Simon and Schuster, 1960.

Scott, Robert Falcon. *The Voyage of the "Discovery."* London: Smith, Elder, 1905.

———. *Scott's Last Expedition, Being the Journals of Captain R. F. Scott, R. N., C. V. O.* Arranged by Leonard Huxley. New York: Dodd, Mead, 1913.

Shackleton, Ernest H. *The Heart of the Antarctic.* New York: Greenwood, 1909.

Shapley, Deborah. *The Seventh Continent: Antarctica in a Resource Age.* Washington: Resources for the Future, 1985.

Siple, Paul A. "Geographical Exploration from Little America III, the West Base of the U. S. Antarctic Service Expedition, 1939–41." *Proceedings of the American Philosophical Society* 89 (1945): 23–60.

Sullivan, Walter. *Quest for a Continent.* New York: McGraw-Hill, 1957.

———. *Assault on the Unknown.* New York: McGraw-Hill, 1961.

Taylor, Griffith. *With Scott: The Silver Lining.* New York: Doss, Mead, 1916.

Vaughan, Norman D. *With Byrd at the Bottom of the World.* Harrisburg: Stackpole, 1990.

Wilson, Edward A. *Diary of the "Discovery" Expedition to the Antarctic Regions, 1901–1904.* Ed. Ann Savours. London: Blandford, 1966.

Yelverton, David E. *Antarctica Unveiled: Scott's First Expedition and the Quest for the Unknown Continent.* Boulder: University of Colorado Press, 2000.

Index